Exploring The BUILDING BLOCKS of Science

Book 6

TEACHER'S MANUAL

REBECCA W. KELLER, PhD

REAL SCIENCE 4 Kids

Exploring the Building Blocks of Science Book 6 Teacher's Manual
ISBN 978-1-941181-15-7

Published by Gravitas Publications Inc.
Real Science-4-Kids®
www.realscience4kids.com
www.gravitaspublications.com

A Note From the Author

This curriculum is designed for middle school level students and provides an introduction to the scientific disciplines of chemistry, biology, physics, geology, and astronomy. *Exploring the Building Blocks of Science Book 6 Laboratory Notebook* accompanies the *Building Blocks of Science Book 6 Student Textbook.* Together, both provide students with basic science concepts needed for developing a solid framework for real science investigation. The *Laboratory Notebook* contains 44 experiments—two experiments for each chapter of the Student Textbook. These experiments allow students to further explore concepts presented in the *Student Textbook.* This teacher's manual will help you guide students through laboratory experiments designed to help them develop the skills needed to use the scientific method.

There are several sections in each chapter of the *Laboratory Notebook*. The section called *Think About It* provides questions to help students develop critical thinking skills and spark their imagination. The *Experiment* section provides students with a framework to explore concepts presented in the *Student Textbook.* In the *Conclusions* section students draw conclusions from the observations they have made during the experiment. A section called *Why?* provides a short explanation of what students may or may not have observed. And finally, in each chapter an additional experiment is presented in *Just For Fun.*

Most of the experiments take up to 1 hour. The materials needed for each experiment are listed on the following pages and also at the beginning of each experiment.

Enjoy!

Rebecca W. Keller, PhD

Materials at a Glance

Experiment 1	Experiment 2	Experiment 3	Experiment 4	Experiment 5
an old digital camera, cell phone, radio, computer, or other small electronic device that is no longer needed small tools such as screwdriver, tweezers, pick rubber gloves, 1-2 pairs library or internet resources chemical glass etching kit glass items for etching (if needed, may be obtained at a thrift store) **Optional** safety goggles	10 ml glass graduated cylinder glass eyedropper 60 ml (1/4 cup) water 60 ml (1/4 cup) rubbing alcohol 60 ml (1/4 cup) vegetable oil waterproofing substance, such as car wax, floor wax, silicone spray, or Scotch-Gard (small amount) additional water and vegetable oil (small amount) **Optional** disposable glass tube Goo Gone or similar cleaner	red cabbage, 1 head distilled water, about 1 liter (1 quart) various solutions, such as: ammonia vinegar clear soda pop milk mineral water large saucepan knife several small jars white coffee filters eyedropper measuring cup measuring spoons marking pen scissors ruler See experiment for list of suggested natural materials for *Just For Fun* section	red cabbage indicator (from Experiment 3) household ammonia vinegar large glass jar measuring spoons measuring cup household solutions chosen by students (to test for acidity and basicity)	tincture of iodine a variety of raw foods, including: pasta bread celery potato banana (ripe) other fruits 1 unripe (green) banana liquid laundry starch (or equal parts borax and corn starch mixed in water) absorbent white paper eye dropper cookie sheet marking pen knife

Experiment 6	Experiment 7	Experiment 8	Experiment 9	Experiment 10
plastic petri dishes* dehydrated agar powder* distilled water K-12 safe *E. coli* bacterial culture* inoculation loop* candle or gas flame cooking pot mixing spoon oven mitt or pot holder measuring spoons measuring cup black permanent marker red marker rubber gloves, 2 pairs * (See experiment for product sources.)	microscope with 4X, 10X, 40X objective lenses. A 100X objective lens is recommended but not required.* glass microscope slides* glass microscope cover slips* immersion oil (if using 100X objective lens)* Samples: piece of paper with lettering strands of hair droplet of blood insect wing *(See experiment for information about how to choose a microscope and for supply sources.)	microscope with a 10X objective* microscope depression slides* 10 or more eyedroppers fresh pond water or water mixed with soil (small amount) protozoa study kit * methyl cellulose* measuring cup and measuring spoons baker's yeast distilled water Eosin Y stain* *(See experiment for product sources.)	dehydrated agar powder* distilled water cooking pot measuring spoons measuring cup plastic petri dishes* permanent marker oven mitt or pot holder jar with lid (big enough to hold 235 ml (about 1 cup) liquid 1 slice of bread, preferably preservative free small clear plastic bag white vinegar bleach borax mold or mildew cleaner 1-2 pairs rubber gloves *(See experiment for product sources.)	One electronic circuit kit (see Experiment 10 for recommendations) **Experiment 11** several glass marbles of different sizes several steel marbles of different sizes cardboard tube, .7-1 meter [2.5-3 ft] long scissors black marking pen ruler letter scale or other small scale or balance

Experiment 12	Experiment 13	Experiment 14	Experiment 15	Experiment 16
stopwatch compass an open space large enough to run (park, schoolyard, playground, backyard, etc.) 5 markers of students' choice to mark distances blank paper a group of friends	pencil or pen marking pen thumbtack or pushpin 3 pieces of string — approximate sizes: 10 cm [4 in.]; 15 cm [6 in.]; 20 cm [8 in.] tape ruler (metric) large piece of white paper (bigger than 30 cm [12 in.] square (can be several pieces of paper taped) firm surface at least as large as the paper and that a thumbtack can be pinned into	pencil, pen, colored pencils compass a small jar or container with a lid small items to place in jar (student selected treasure) garden trowel (optional)	computer with internet access (a program that unzips files may be needed) **Optional** printer and paper colored pencils	Some suggestions for student chosen model making materials: modeling clay of different colors marble or steel ball ingredients to make various colored cakes materials for making paper mache Styrofoam balls

Experiment 17	Experiment 18	Experiment 20	Experiment 21	Experiment 22
2 liter (2 quart) plastic bottle with cap warm water matches blank paper	two sticks (used for marking locations) two rulers tape string, several meters long (several yards) protractor **Experiment 19** computer with internet access printer and paper flashlight **Optional** binoculars or telescope star map app and mobile device	8 objects of different sizes to represent the planets ruler (in centimeters) marking pen large flat surface for drawing — 1 x 1 meter (3 x 3 feet), such as a large piece of cardboard or several sheets of construction paper large open space at least 3 meters (10 feet) square push pin piece of string one meter (3 feet) long additional objects of students' choice	pencil colored pencils **Optional** blank paper, several sheets	computer with internet access materials as needed for project chosen by students blank paper or notebook

Materials
Quantities Needed for All Experiments

Equipment	Foods
bottle, plastic, 2 liter (2 quart) with cap compass computer with internet access cookie sheet electronic circuit kit (see Chapter 10 for recommendations) electronic device, old-unneeded: digital camera, cell phone, radio, computer, or other small electronic device flashlight graduated cylinder, glass, 10 ml inoculation loop[1] knife marbles, glass, several different sizes marbles, steel, several different sizes measuring cup measuring spoons microscope with 4X, 10X, 40X objective lenses. A 100X objective lens is recommended but not required.[2] natural materials for *Just For Fun* section (see Exper. 2) oven mitt or pot holder pot, cooking printer and paper or mobile device protractor ruler (in centimeters) rulers, 2 saucepan, large scale: letter, or other small scale or balance scissors spoon, mixing stopwatch tools, small - such as screwdriver, tweezers, pick **Optional** binoculars or telescope mobile device safety goggles trowel, garden	baker's yeast banana, 1 unripe (green) bread, 1 slice, preferably preservative free cabbage, red, 1 head foods, raw-including: pasta, bread, celery, potato, banana (ripe), misc. fruits vegetable oil, somewhat more than 60 ml (1/4 cup) vinegar vinegar, white water

[1] See Experiment 6 for supply sources.
[2] See Experiment 7 for supply sources.
[3] See Experiment 8 for supply sources.

Materials

Quantities Needed for All Experiments

Materials

agar powder, dehydrated[1]
ammonia, household
bag, clear plastic, small
bleach
blood, 1 droplet
borax
cleaner, mold or mildew
coffee filters, white
E. coli bacterial culture, K-12 safe[1]
Eosin Y stain[3]
eyedropper, glass, 1-2 dozen
glass etching kit, chemical
glass items for etching (if needed, may be obtained at a thrift store)
gloves, 4-6 pairs, rubber
hair, several strands
immersion oil (if using 100X objective lens)[2]
insect wing
iodine, tincture of
items, misc. small, to place in jar (student selected treasure)
jar with lid (big enough to hold 235 ml (about 1 cup liquid)
jar, glass, large
jar, small with lid, or small container with lid
jars, small, several
marker, black permanent
marker, red
markers, 5 items of students' choice to mark distances
matches
materials as needed for project chosen by students
methyl cellulose[3]
microscope cover slips, glass[2]
microscope slides, glass, depression[3]
microscope slides, glass, regular[2]
objects of students' choice
objects, 8 of different sizes to represent the planets
paper with lettering, small piece
paper, absorbent white

Materials (continued)

paper, blank
paper, blank, or notebook
paper, white (bigger than 30 cm [12 in.] square (can be several pieces of paper taped)
pen
pen, black marking
pen, marking
pencil
pencils, colored
petri dishes, plastic[1]
protozoa study kit[3]
pushpin or thumbtack
pushpin, 1-2
rubbing alcohol, 60 ml (1/4 cup)
solutions, household, chosen by students (to test for acidity and basicity)
solutions,various-such as: ammonia, vinegar, clear soda pop, milk, mineral water
starch, liquid laundry (or equal parts borax and corn starch mixed in water)
sticks, 2 (used for marking locations)
string, one meter (3 feet) long
string, several meters long (several yards)
string, 3 pieces— approximate sizes: 10 cm [4 in.]; 15 cm [6 in.]; 20 cm [8 in.]
tape
tube, cardboard, .7-1 meter [2.5-3 ft] long
water, distilled, 2 liters or more
water, fresh pond water or water mixed with soil (small amount)
waterproofing substance, such as car wax, floor wax, silicone spray, or Scotch-Gard (small amount)
student chosen model making materials, such as:
clay, modeling, different colors
ball, steel, or marble
ingredients for various colored cakes
paper mache materials
balls, Styrofoam

Optional

Goo Gone or similar cleaner
tube, glass, disposable

Other

computer program that unzips files (may be needed)
flame, candle or gas
friends, several
open space at least 3 meters (10 feet) square
open space large enough to run (park, schoolyard, playground, backyard, etc.)
resources, library or internet
surface, firm, at least 30 cm [12 in.] square that a thumbtack can be pinned into
surface, large, flat, for drawing: 1 x 1 meter (3 x 3 feet), such as a large piece of cardboard or several sheets of construction paper

Optional

star map app (Exper. 19)

[1] See Experiment 6 for supply sources.
[2] See Experiment 7 for supply sources.
[3] See Experiment 8 for supply sources.

Contents

Experiment 1

Take It Apart!

Materials Needed

- an old digital camera, cell phone, radio, computer, or other small electronic device that is no longer needed
- small tools such as screwdriver, tweezers, pick
- rubber gloves, 1-2 pairs
- library or internet resources
- chemical glass etching kit
- glass items for etching (if needed, may be obtained at a thrift store)

Optional

- safety goggles

Objectives

In this experiment students will explore how a modern device like a cell phone, camera, or radio is a combination of both technological advances and scientific discoveries.

The objectives of this lesson are for students to:

- Observe the connection between science and technology.
- Learn how chemical etching (a scientific discovery) is used to make printed circuit boards (a technological advance).

Experiment

I. Think About It

Read this section of the *Laboratory Notebook* with your students.

Explore open inquiry with questions such as the following.

- *Do you think a cell phone and a digital camera (or other electronic devices) are made of the same materials? Why or why not?*
- *What materials do you think they are made of?*
- *If you opened the cases of a cell phone and a digital camera, do you think they would both look the same on the inside? Why or why not?*
- *Do you think a camera, cell phone, radio, television set, and computer are similar? Why or why not?*
- *In what ways do you think a camera, cell phone, radio, television set, and computer are different? Why?*
- *Because a camera, cell phone, radio, television set, and computer are used for different purposes, do you think they need to be made of different materials? Different parts? Why or why not? What differences would they need to have?*

II. Experiment 1: Take It Apart!

Have the students read the entire experiment before writing an objective and a hypothesis.

Objective: Have the students write an objective for this experiment (what they think they will be learning).

Hypothesis: Have the students write a hypothesis. The hypothesis can restate the objective in the form of a statement that can be proved or disproved by their experiment. Some examples include:

- *My old cell phone is made of both plastic and metal.*
- *My old radio contains metal and plastic wires.*
- *My small electronic toy is powered by batteries.*
- *My old digital camera contains an electronic circuit board.*

EXPERIMENT

❶ Help the students find an old electronic device that can be easily taken apart. It will most likely be destroyed in the process. If you don't have a device available, look for one at a thrift store or junk yard.

❷ Have the students observe the outside of the device and note what materials the device is made of—plastic, metal (what kind?), glass, etc. In the space provided, have them record their observations.

❸ Have the students put on rubber gloves. With your supervision, have them use the appropriate tools to carefully disassemble the device, taking off the back or outer covering and observing the inside. Have them take out or pry off any parts that can be removed. Have them work carefully so no parts of the device break violently. **Do not let them pry open or destroy batteries.** You may want to have them wear goggles to protect their eyes from any flying parts.

In the space provided, have the students draw the parts they've removed and make notes about them.

Results

❶ Have the students use the library or internet to research how the device was made, where it was made, what it is made of, and who designed it. Have the students check the device and components for logos or brands that might identify the source of the components.

Have the students record the results of their research.

❷ Have the students think about how the device was made. Using the library or an internet resource explore with your students the scientific processes involved in making the device. Look for ways chemistry, physics, biology, geology, and/or astronomy were involved in the processes of designing, crafting, and assembling the device. Have the students record their observations.

III. Conclusions

Have the students review the observations they recorded for the experiment. Have them draw a conclusion based on their observations and research. Have them note if their conclusion supports or does not support their hypothesis.

IV. Why?

Read this section of the *Laboratory Notebook* with your students.
Discuss any questions that might come up.

Most modern electronic devices contain a printed circuit board. Discuss with the students how PCBs could not have been developed without chemistry, including the process of chemical etching.

V. Just For Fun

Chemical etching can be used on a variety of different materials. Help your students find a simple glass etching kit that will help them observe how this process works. Have them write and draw their ideas of what designs they might create with etching. If needed, glass items can be purchased at a thrift store. You may want to have students wear rubber gloves and goggles while doing the etching.

Experiment 2

Reading the Meniscus

Materials Needed

- 10 ml glass graduated cylinder
- glass eyedropper
- 60 ml (1/4 cup) water
- 60 ml (1/4 cup) rubbing alcohol
- 60 ml (1/4 cup) vegetable oil
- waterproofing substance, such as car wax, floor wax, silicone spray, or Scotch-Gard (small amount)
- additional water and vegetable oil (small amount)

Optional

- disposable glass tube
- Goo Gone or similar cleaner

Objectives

In this experiment students will explore how to correctly read a graduated cylinder.

The objectives of this lesson are for students to:

- Practice using chemistry lab equipment.
- Explore how different liquids behave in glass lab equipment.

Experiment

I. Think About It

Read this section of the *Laboratory Notebook* with your students.

Ask questions such as the following to guide open inquiry. After the students have discussed their ideas, they can experiment with placing a droplet of water and a droplet of oil on various surfaces to observe what happens.

- *If you put a droplet of water on a plastic surface, what do you think will happen?*
- *If you put a droplet of water on a glass surface, what do you think will happen?*
- *If you put a droplet of oil on a plastic surface, what do you think will happen?*
- *If you put a droplet of oil on a glass surface, what do you think will happen?*
- *Why do you think water spreads out on a glass surface and oil does not?*

II. Experiment 2: Reading the Meniscus

Have the students read the entire experiment before writing an objective and a hypothesis.

Objective: Have the students write an objective. Some examples:

- *To find out how oil and water will behave in a glass graduated cylinder.*
- *To learn how to read the volume of water in a glass graduated cylinder.*

Hypothesis: Have the students write a hypothesis (what they think they will be learning). The hypothesis can restate the objective as a statement that can be proved or disproved by their experiment. Some examples include:

- *Oil and water will behave in the same way in a glass graduated cylinder.*
- *Oil and water will behave differently in a glass graduated cylinder.*
- *It will be easy to read the volume of water in a glass graduated cylinder.*
- *It will be difficult to read the volume of water in a glass graduated cylinder.*

EXPERIMENT

❶ Have the students observe the details of the graduated cylinder. Have them note the width of the mouth, the pour spout, and the markings along the side.

❷-❸ Have the students pour 5 ml of water into the graduated cylinder and then, holding the graduated cylinder or placing it on a table, have them align their eyes with the top surface of the water. It is expected that they will notice that the water level is higher where the water is against the glass than it is in the center, creating a curve. Have the students record their observations in the chart provided in the *Results* section.

❹ Explain that the curvature of the surface of the water is called the *meniscus*. Water in a glass graduated cylinder will form a concave meniscus, with the surface of the water curving downward to the center from where the water is against the glass. Oil in a glass graduated cylinder will form a convex meniscus, with the surface of the water curving upward from the glass to the center.

Have the students observe whether the bottom of the meniscus is above or below the 5 ml mark on the graduated cylinder.

❺ Students will now need to adjust the water level until the bottom of the meniscus aligns exactly with the 5 ml mark. This may be difficult at first. It is easy to pour too much liquid in and then after pouring some out, find that too much has been poured out. Have the students use the eyedropper to add small amounts of water until the 5 ml mark aligns with the bottom of the curvature of the water.

❻-❼ Have the students repeat Steps ❷-❺ with rubbing alcohol and then with vegetable oil in that order.

Liquids with a concave meniscus are measured at the bottom of the meniscus, and liquids with a convex meniscus are measured at the top of the meniscus. In both cases the level of the liquid is being measured by its height at the center of the cylinder.

Results

A chart is provided for students to record their observations.

III. Conclusions

Have the students draw a conclusion based on their observations and research. Have them note whether their conclusion supports or does not support their hypothesis.

IV. Why?

Read this section of the *Laboratory Notebook* with your students.
Discuss any questions that might come up.

Discuss how liquids will be either attracted to or repelled by different surfaces and how this affects reading measurements in volumetric glassware, such as a graduated cylinder.

V. Just For Fun

Have the students repeat the experiment by applying a water repellent material, such as liquid car wax, floor wax, silicone spray, or Scotch-Gard, to the inside of the graduated cylinder or a disposable glass tube. To remove the water repellent from the graduated cylinder after the experiment, soak it in a hydrophobic cleaner such as Goo Gone and then wash with soap and water.

Have the students pour water into the graduated cylinder, observe the meniscus, record their results, and then repeat with vegetable oil.

A chart is provided for students to record their observations of the meniscuses formed by the liquids in the waxed cylinder and to compare the results of the *Reading the Meniscus* experiment and the *Just For Fun* experiment.

Experiment 3

Making an Acid-Base Indicator

Materials Needed

- one head of red cabbage
- distilled water, about 1 liter (1 quart)
- various solutions, such as:
  ammonia
  vinegar
  clear soda pop
  milk
  mineral water
- large saucepan
- knife
- several small jars
- white coffee filters
- eyedropper
- measuring cup
- measuring spoons
- marking pen
- scissors
- ruler

Suggested Natural Materials for Just For Fun (help students select several)

- Turmeric
- Poppyseed or cornflower petals
- Madder plant (Rubiaceae family)
- Red beets
- Rose petals
- Berries
- Blue and red grapes
- Cherries
- Geranium petals
- Morning glory
- Red onion
- Petunia petals
- Hibiscus petals (or hibiscus tea)
- Carrots
- Other strongly colored plant materials of students' choice

Objectives

In this experiment students will be introduced to the concepts of acids, bases, pH, and pH indicators.

The objectives of this lesson are for students to:

- Observe that acids and bases have different properties that can be tested for.
- Use controls in an experiment.

Experiment

I. Think About It

Read this section of the *Laboratory Notebook* with your students.

Ask questions such as the following to guide open inquiry.

- *What do you think an acid is?*
- *What liquids can you think of that are acids?*
- *What do you think a base is?*
- *What liquids can you think of that are bases?*
- *How would you find out if a solution is an acid or a base?*
- *Do you think you use acids and bases in your everyday life? Why or why not?*

II. Experiment 3: Making an Acid-Base Indicator

Have the students read the entire experiment before writing an objective and a hypothesis.

Objective: Have the students write an objective. An example:

- *We will make an acid-base indicator from red cabbage and use it to determine whether solutions are acidic or basic.*

Hypothesis: Have the students write a hypothesis. An example:

- *We can use an indicator to identify solutions as acidic or basic.*

EXPERIMENT

In this experiment the students will use "controls." A control is an experiment where the outcome is already known or where a given outcome can be determined. The control provides a point of reference or comparison for experiments that use unknowns. For example, in this experiment the students will test for acidity or basicity with a pH indicator, but they do not know what the expected color change will be. By doing controls with solutions that they know are either acidic (vinegar) or basic (ammonia), they can determine what the color change for an acid is and what the color change for a base is. Only then can they test the "unknown" solutions.

The liquids in the materials list include both acids and bases. Milk in neutral. Have the students be careful when handling ammonia. Other suggested items to test include:

- water (neutral)
- Windex or other glass cleaner (basic)
- Lemon juice or orange juice (acidic)
- White grape juice (acidic)

❶-❸ Have the students prepare red cabbage juice indicator. Have them cut a head of red cabbage into several pieces, put the pieces in about .7 liter (3 cups) of boiling distilled water, and boil the cabbage for several minutes until the liquid is a deep purple color. Have them remove the cabbage and let the water cool.

❹ Have the students set aside .25 liter (1 cup) of the red cabbage juice and refrigerate the rest to use in the next experiment. It is important to refrigerate the cabbage juice or it will spoil and cannot be used for the next experiment. It should keep about two weeks in the refrigerator.

❺ Have the students cut 20 or more strips of coffee filter paper that are about 2 cm (3/4 in.) by 4 cm (1½ in.) for testing the solutions.

❻ Students will make pH paper by using an eyedropper to put several drops of the red cabbage juice on each strip of coffee filter paper and letting it dry. If the cabbage indicator is added to the strips of paper several times and dried in between, the color change when testing the liquids will be more dramatic.

❼ Have the students label one jar *Control Acid*. They will make the control acid by putting in the jar 15 ml (1 tbsp.) of vinegar and 75 ml (5 tbsp.) of distilled water.

A second jar will be labeled *Control Base*. Students will make the control base by putting in the jar 15 ml (1 tbsp.) of ammonia and 75 ml (5 tbsp.) of distilled water.

Have the students label a jar for each of the solutions they will be testing and then put 15 ml (1 tbsp.) of each substance in the appropriate jar along with 30-75 ml (2-5 tbsp.) of distilled water.

❽-❾ Students will test the *Control Acid* and the *Control Base* by dipping an unused strip of pH paper into each. Have them record their results and tape the pH papers into their *Laboratory Notebook* in the chart in the *Results* section.

The vinegar (control acid) should turn the paper pink.
The ammonia (control base) should turn the paper green.

The color change of the pH paper may be quick and noticeable or it may be subtle. It is best to have the students look at the paper immediately after it has been dipped into the solution. If it is too difficult to determine the color change of the paper, the cabbage indicator can be used directly in the solution. Have the students pour a small amount (5-10 ml [1-2 teaspoons]) into the solution and record the color change.

❿ Have the students test the remaining solutions and record their results.

III. Conclusions

Have the students review the results they recorded for this experiment. Have them draw conclusions based on the data they collected. Help the students to be specific and to make valid conclusions from their data. If a solution did not change color, but the experimental controls worked, it is probably true that the solution is neutral or near neutral. However, if no color change is observed or if the result is ambiguous, it may not be true that the solution is neutral, and it may be true that the color change is too subtle to be easily perceived.

Have the students draw conclusions even if they experienced difficulties with the experiment.

IV. Why?

Read this section of the *Laboratory Notebook* with your students.
Discuss any questions that might come up.

V. Just For Fun

Students will test different natural materials to see if they are acid-base indicators.

Help the students select and gather the materials to be tested. Have them use several materials from the list provided, and they can also try other natural materials that have a strong color. Students can use their control acid and base or other solutions that they have identified as acidic or basic.

Have the students crush (or chop) the material to be tested and put some of it in each of two small jars. They then will add an acid to one jar and a base to the other and note whether there is a color change. If there is no color change, you can direct them to make a stronger solution of the acid and base they have chosen and see if this makes a difference. They can also experiment with adding a little distilled water to the material being tested before adding the acid or base. In the chart provided, have the students record their results including whether or not they think the material is an acid-base indicator.

Some examples:

- Turmeric powder will turn red in a base (will be yellow at pH 7.4 and red at pH 8.6).
- Poppyseed or cornflower petals contain the same chemical as red cabbage and will undergo a similar color change.
- Madder plant (Rubiaceae) will turn from yellow to red in a weak base (yellow at pH 5.5 and red at pH 6.8).
- Cherries and cherry juice turn from red to purple in a base.

Additional Notes

An acid-base reaction is a type of exchange reaction. In the example illustrated in this chapter of the textbook, the molecules are not drawn with the bonds showing, and on first inspection it appears that the central carbon of both molecules has broken the rule of "4 bonds for carbon." Also, two of the oxygens appear to have broken the rule for "2 bonds for oxygen." However, in each case the bond between the central carbon atom and one of the oxygen atoms is a double bond. Double bonds are beyond the scope of this level, but all of the bonding rules are satisfied.

pH is actually a measure of the hydrogen ion concentration (written as [H]).The pH scale is important, but mathematically and conceptually the actual definition of pH is too difficult for this level. (The mathematical expression for pH is: $pH = -\log [H]$)

The higher the hydrogen ion concentration, the lower the pH; the lower the hydrogen ion concentration, the higher the pH. The hydrogen ion concentration is the real definition of what is meant by "acid" in this chapter.

Scientists measure pH with pH meters, pH paper, or solution indicators, with the use of the pH meter being the most common laboratory technique. There are a variety of pH meters and electrodes available. The most common electrode is called a glass electrode. There is a small glass ball at the end of this electrode that senses the pH electrically.

Before pH meters, pH paper was the most common way to measure pH. Litmus paper can still be found in most laboratories along with other types of pH paper. Litmus paper is made with a compound called an indicator. An indicator is any molecule that changes color as a result of a pH change.

There are two types of litmus paper—blue litmus paper tests for acidic solutions, and red litmus paper tests for basic solutions. Litmus paper is not suitable for determining the exact pH; it can only indicate whether a solution is acidic or basic. Other types of pH paper can more accurately determine the actual pH.

The chart in this section of the *Student Textbook* shows some common indicators used in the laboratory and is meant to illustrate that there are a variety of pH indicators that can be used over a wide range of pH. Often pH indicators are mixed so that more than one pH range can be detected. The names of some of these indicators are difficult to pronounce, but many of them can be looked up in a dictionary, encyclopedia, or online for pronunciation guidance.

Experiment 4

Vinegar and Ammonia in the Balance: An Introduction to Titration

Materials Needed

- red cabbage juice indicator (from Experiment 3)
- household ammonia
- vinegar
- large glass jar
- measuring spoons
- measuring cup
- household solutions chosen by students (to test for acidity and basicity)

Objectives

In this experiment students will explore acid-base reactions by performing a titration.

The objectives of this lesson are for students to:

- Perform a titration.
- Plot and analyze data on a graph.

Experiment

I. Think About It

Read this section of the *Laboratory Notebook* with your students.

Ask questions such as the following to guide open inquiry.

- *What do you think happens when you mix an acid and a base together? Why?*
- *What products do you think you would get by mixing an acid and a base together?*
- *Do you think knowing the concentration of an acid or a base could be helpful? Why or why not?*
- *Do you think there are times when you would want to have a solution that is neutral? Why or why not?*
- *Do you think plotting data on a graph can be useful? Why or why not?*

II. Experiment 4: Vinegar and Ammonia in the Balance: An Introduction to Titration

Have the students read the entire experiment before writing an objective and a hypothesis.

Objective: Have the students write an objective. For example:

- *To determine how much ammonia is needed to change the color of red cabbage juice indicator in vinegar from red to green*

Hypothesis: Have the students write a hypothesis. For example:

- *An indicator can be used to observe the acid-base reaction of vinegar and ammonia.*

EXPERIMENT

In this experiment, students will perform an acid-base titration using a red cabbage juice indicator. The red cabbage juice indicator from Experiment 3 is required. If the indicator is too old (more than two weeks or has mold or bacteria growing in it), have the students make fresh cabbage indicator.

NOTE: This titration can be tricky if the concentration of the base is too dilute.

A quick test can be performed by the teacher without the students' observation. Take 60 ml (1/4 cup) of vinegar and add indicator to it. See that it turns red. Add 60 ml (1/4 cup) ammonia directly to this acid-indicator mixture. The color should turn green, but if the color is still red, add another 60 ml (1/4 cup) of ammonia. It should turn green; however, if it does not, dilute the vinegar with 120 ml (1/2 cup) water and repeat the above steps.

This quick "titration" will help determine how much total ammonia is needed to neutralize the acid. Adjust the titration so that not much more than 60 ml (1/4 cup) of ammonia is needed. Less is all right, but the students will get frustrated if they have to add more than 100 ml (20 teaspoons) of ammonia, and the best part of the titration is the last part when they see the color change occur.

❶-❷ Have the students measure 60 ml (1/4 cup) of vinegar into a jar and add enough red cabbage juice to get a deep red color.

❸-❻ Students will add ammonia to the vinegar 5 ml (1 tsp.) at a time. Each time the students add ammonia, they will swirl the solution and then record the color of the solution and the total amount of ammonia that has been added. There is a chart provided in the **Results** section.

As the students add ammonia, the color stays mostly red, then turns a little purple, and finally turns all green. The transition is quite striking. Have the students continue adding ammonia to see that the color stays green.

Graphing Your Data

❶-❸ Have the students take the data from their chart and plot it on the graph provided. When all the data has been plotted, have them connect the data points.

The data points should look something like those shown in the following graph. Many points lie along the bottom left of the plot, then one or two points will be in the middle. Finally, several will be along the top right-hand side of the plot.

Have the students connect the points with a smooth curved line. Their plot should look similar to the following. Discuss the following parts of the graph:

- In the left-hand lower portion of the plot, the solution is acidic.
- In the middle portion, where the line is going upward, the solution is between acidic and basic (near neutral).
- In the upper right-hand corner, the solution is basic.

Point out that we know this because the color of the indicator is known at various pH values, as we observed in Experiment 3.

Example *(Answers may vary.)*

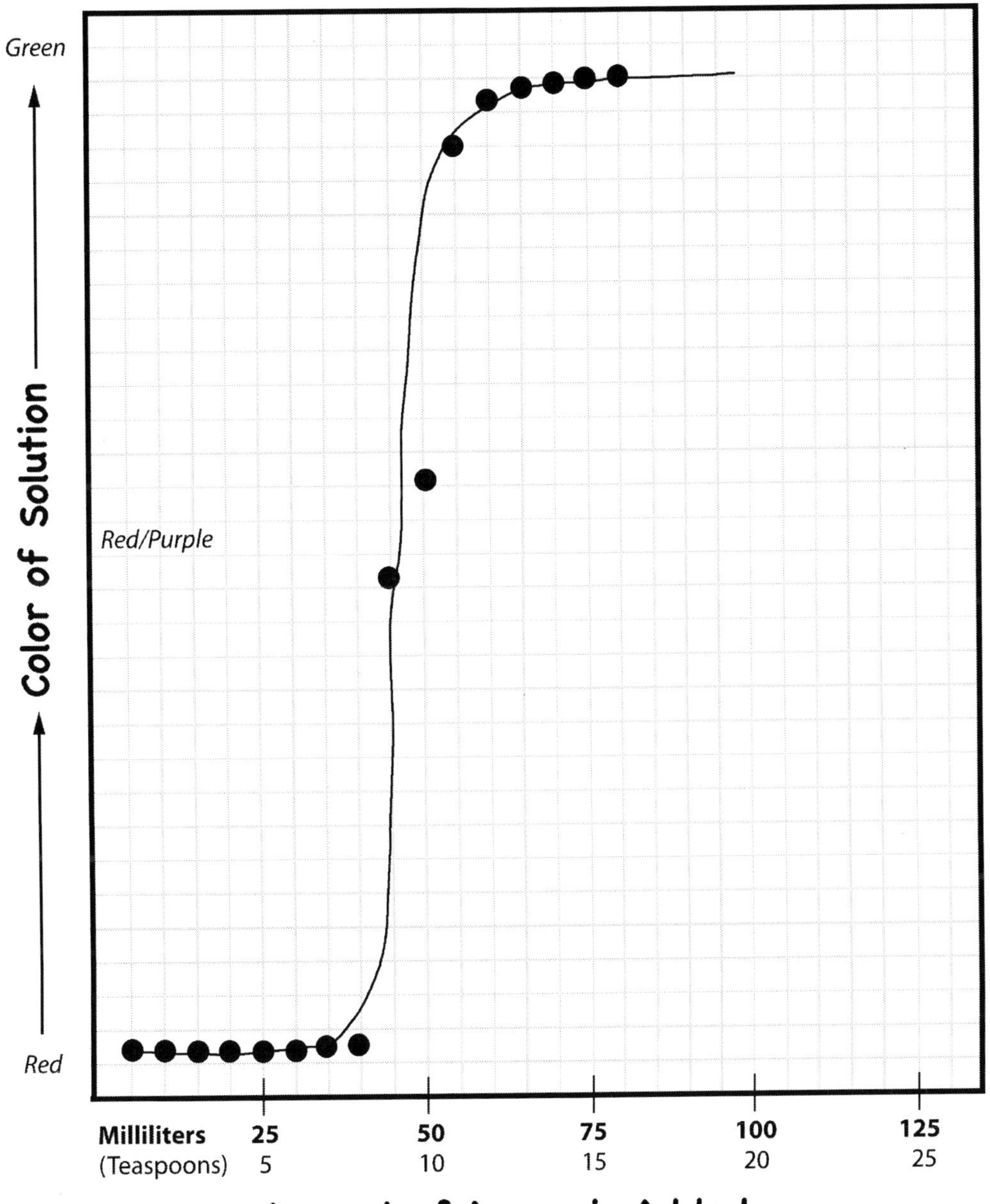

CHEMISTRY

III. Conclusions

Have the students review the results they recorded for the experiment. Have them draw conclusions based on the data they collected.

Have the students write down the amount of ammonia it took to neutralize the vinegar. Note whether this equals the amount of vinegar that was used—60 ml (1/4 cup [12 tsp.]). Depending on the brands of vinegar and ammonia used, the amounts are often equal.

Help them reach valid conclusions. Some examples:

- *It took 55 ml of ammonia to turn the solution green.*
- *It took 11 teaspoons to neutralize the vinegar with ammonia.*
- *The amount of ammonia required to neutralize the vinegar was equal to the amount of vinegar used.*

IV. Why?

Read this section of the *Laboratory Notebook* with your students. Discuss any questions that might come up.

V. Just For Fun

Students are to repeat the experiment with a different acid and base.

Help the students find household solutions that they think may be acidic or basic. Have them use red cabbage juice indicator to test each solution and then select one acid and one base. Have the students follow the steps of the previous experiment to do a titration.

Some possible acid/base combinations to test:

- Clear soda pop (acid) and baking soda water (base).
- Clear soda pop (acid) and ammonia (base).
- Dilute clear window cleaner (base) with an acid like vinegar or clear soda pop.
- Lye [sodium hydroxide] (base) with an acid like vinegar or clear soda pop. **(Lye is caustic—adult supervision is required.)**

A chart is provided for the collection of data and a graph for plotting the data.

Experiment 5

Show Me the Starch!

Materials Needed

- tincture of iodine **[Iodine is VERY poisonous—DO NOT LET STUDENTS EAT any food items with iodine on them.]**
- a variety of raw foods, including:
  pasta
  bread
  celery
  potato
  banana (ripe) and other fruits
- 1 unripe (green) banana
- liquid laundry starch (or equal parts borax and corn starch mixed in water)
- absorbent white paper
- eye dropper
- cookie sheet
- marking pen
- knife

Objectives

In this experiment students will be introduced to "energy" molecules such as carbohydrates and starches that fuel our bodies.

The objectives of this lesson are for students to:

- Perform a simple test for starches.
- Use a control sample for comparison.

Experiment

I. Think About It

Read this section of the *Laboratory Notebook* with your students.

Ask questions such as the following to guide open inquiry.

- *Why do you think your body needs carbohydrates?*
- *What foods do you think contain carbohydrates? Why?*
- *What do you think carbohydrates are made of?*
- *Do you think all sugars are the same? Why or why not?*
- *Do you think there are different kinds of starches? Why or why not?*
- *Do you think grass would be good for you to eat? Why or why not?*

II. Experiment 5: Show Me the Starch!

Have the students read the entire experiment before writing an objective and a hypothesis.

Objective: Have the students write an objective. For example:

- *To determine which foods contain starch.*

Hypothesis: Have the students write a hypothesis. For example:

- *Potatoes contain starch. Celery does not.*

EXPERIMENT

❶ Have the students select food items to test and place them on a cookie sheet.

❷ Students will make a *control* by using an eye dropper to put a small amount of laundry starch (or a borax/cornstarch/water mixture) on a piece of absorbent paper and letting it dry before testing it.

❸ Have the students add a drop of iodine to the starch on the control paper and record the results in the chart provided in the *Results* section. **[Iodine is VERY poisonous — DO NOT LET STUDENTS EAT any food items with iodine on them.]**

❹ Have the students add a drop of iodine to each of the food samples and record their results.

❺-❻ Have the students compare the color of the *Control* to each of the food items tested and note which foods turned color.

Results

A chart is provided for recording the results of the experiment.

III. Conclusions

Have the students review the results they recorded for the experiment. Have them draw conclusions based on the data they collected.

IV. Why?

Read this section of the *Laboratory Notebook* with your students.
Discuss any questions that might come up.

V. Just For Fun

Students will observe how a banana changes as it ripens. As the banana ripens, the carbohydrates start to break apart, releasing small sugar molecules that give the banana its sweet taste.

Have the students begin with a green banana. Have them slice off a piece of the banana and test it with a drop of iodine. Again instruct them not to eat any food with iodine on it. Have them leave the banana at room temperature, and every third day have them test another slice of the banana. In the chart provided, have the students record their results after each test. When they have completed the experiment, have them review what happened over the course of the experiment.

Experiment 6

Using Agar Plates

Materials Needed

- plastic petri dishes*
- dehydrated agar powder**
- distilled water
- K-12 safe *E. coli* bacterial culture http://www.hometrainingtools.com/escherichia-coli-bacteria/p/LD-ESCHCOL/
- inoculation loop http://www.hometrainingtools.com/inoculating-needle-looped-end/p/BE-INOCUL/
- candle or gas flame
- cooking pot
- mixing spoon
- oven mitt or pot holder
- measuring spoons
- measuring cup
- black permanent marker
- red marker
- rubber gloves, 2 pairs

* A stack of 20 can be ordered from: http://www.hometrainingtools.com/petri-dishes-plastic-20-pk/p/BE-PETRI20/

** http://www.hometrainingtools.com/nutrient-agar-8-g-dehydrated/p/CH-AGARN08/

Objectives

In this experiment students will explore how to prepare and use agar plates.

The objectives of this lesson are for students to:

- Practice using microbiology and genetics lab equipment.
- Explore how agar plates may differ, how to observe bubbles in the agar, and what happens when cultures are spread vs. streaked.

Experiment

I. Think About It

Read this section of the *Laboratory Notebook* with your students.

Ask questions such as the following to guide open inquiry.

- *Do you think one agar plate can differ from another? Why or why not?*
- *Do you think agar plates can differ in quality? Why or why not? How?*
- *Do you think the quality of an agar plate matters? Why or why not?*
- *Do you think bubbles or other defects in an agar plate might cause problems for bacterial growth? Why or why not?*
- *Do you think water condensing on an agar plate might create problems for bacterial growth? Why or why not?*
- *What do you think will happen if an agar plate dries out? Will bacteria grow? Why or why not?*

II. Experiment 6: Using Agar Plates

Have the students read the entire experiment before writing an objective and a hypothesis.

Objective: Have the students think of an objective for this experiment (what will they be learning?).

Hypothesis: Have the students write a hypothesis. The hypothesis can restate the objective in a statement that can be proved or disproved by their experiment. Some examples include:

- *Bubbles will not affect bacterial growth on an agar plate.*
- *Bubbles will make it hard to detect bacterial growth.*
- *Water condensation will damage the surface of an agar plate.*
- *Bacteria will not grow on a dry agar plate.*
- *Bacteria will grow the same way on all the plates.*

EXPERIMENT

Part I: Preparing Agar Plates

Help the students assemble the materials for making agar plates. Keep the experimental area as clean as possible to avoid contamination.

❶ Have the students prepare a clean, flat surface on which to pour the plates. Have them spread out 18-20 petri dishes to prepare for both the *Using Agar Plates* experiment and the *Just For Fun* experiment.

❷ Have the students add 10 ml (2 teaspoons) dehydrated agar powder to 200 ml (about 1 cup) room temperature distilled water in a cooking pot and bring to a boil while stirring.

❸ The hot agar can be cooled slightly before pouring but should not cool so much that it starts to harden. Have the students use an oven mitt or pot holder while picking up the pot of agar. To prevent contamination, have them slide the lid partially off a petri dish just before they fill it and then re-cover it immediately after it is filled. Have them carefully pour a small amount of the hot agar into each of the petri dishes. It is easy to pour too much agar in one petri dish and too little in another, but they should try to put just enough agar in each petri dish to cover the bottom.

Students will need to have one or more plates that have bubbles in the agar. Each plate that contains bubbles should be marked with a red dot on the lid. If no bubbles are forming on the plates, have the students shake the pot of hot agar before pouring the last few plates.

❹ Once all the plates have been poured, they should be allowed to cool until the agar forms a hard surface.

❺ Have the students mark the lids of four plates as follows:

Take one of the plates that has bubbles and turn it upside down, agar side up. Mark it "Bubbles" and store it agar side up in the refrigerator.

Take a plate without bubbles and turn it upside down, agar side up. Mark it "Up" and store it agar side up in the refrigerator.

Take a plate without bubbles and leave it agar side down. Mark it "Down" and store it agar side down in the refrigerator.

Take a plate with or without bubbles, remove the lid, and allow it to dry until the edges of the agar start to pull away from the sides of the petri dish. This may take a few days depending on the humidity in your region. Replace the lid and mark it "Dry." Store right side up at room temperature.

❻ Have the students store the remaining plates upside down (agar side up) in the refrigerator.

Part II: Streaking the *E. coli* Culture

Have the students wear rubber gloves during the experiment. Explain that scientists in a laboratory wear rubber gloves to prevent contamination of samples.

❶-❷ Have the students sterilize the inoculating loop by heating it in a gas flame or in a candle flame until the wire turns orange and then let it cool without touching anything. [Note: If using a candle flame, the loop should be kept close to the blue part of flame to avoid soot.]

❸ Have the students dip the loop into the tube of *E. coli* culture. They need to be careful not to let the loop touch the sides of the tube as they insert and remove it. Repeat for each plate.

❹ Have the students streak the marked plates with the inoculation loop in a zigzag pattern from one side of the petri dish to the other. The inoculated plates will then be stored agar side up at room temperature.

❺ Using one agar plate that has no bubbles and was stored in the refrigerator agar side up, the students are to pour a small volume of *E. coli* culture directly on the agar (enough to cover the agar) and then allow the agar to absorb the culture. This plate will be marked "Spread," then stored agar side up at room temperature with the other plates.

Results

Have the students examine the plates carefully after they begin to observe growth (about 1-3 days). Have them record their observations in the chart provided.

III. Conclusions

Have the students draw conclusions based on their observations and research. What differences did they see between the plates that were prepared in different ways?

IV. Why?

Discuss why preparing good agar plates is important for performing good microbiological and genetics experiments. Bacterial growth may not occur properly if the agar plates have bubbles, condensation, or are dried out.

V. Just For Fun

Have students practice different streaking patterns and then, in about 1-3 days, observe how well or poorly these streaks result in the production of single colonies.

Experiment 7

Using a Light Microscope

Materials Needed

- microscope with 4X, 10X, and 40X objective lenses. A 100X objective lens is recommended but not required.
- glass microscope slides[1]
- glass microscope cover slips[2]
- immersion oil (if using 100X objective lens)[3]
- Samples:
  piece of paper with lettering
  strands of hair
  droplet of blood
  insect wing

Suggested source:
http://www.hometrainingtools.com/

[1] glass microscope slides
MS-SLIDP72 or MS-SLIDEPL

[2] glass microscope cover slip
MS-SLIDCV

[3] immersion oil
MI-IMMOIL

How to Buy a Microscope

What to Look For

- A metal mechanical stage.
- A metal body painted with a resistant finish.
- DIN Achromatic Glass objective lenses at 4X, 10X, 40X (a 100X lens is optional but recommended).
- A focusable condenser (lens that focuses the light on the sample).
- Metal gears and screws with ball bearings for movable parts.
- Monocular (single tube) "wide field" ocular lens.
- Fluorescent lighting with an iris diaphragm.

Price Range

$50-$150: Not recommended: These microscopes do not have the best construction or parts and are often made of plastic. These microscopes will cause frustration, discouraging students.

$150-$350: A good quality standard student microscope can be found in this price range. We recommend Great Scopes for a solid student microscope with the best parts and optics in this price range. http://www.greatscopes.com

Above $350: There are many higher end microscopes that can be purchased, but for most students these are too much microscope for their needs. However, if you have a child who is really interested in microscopy, wants to enter the medical or scientific profession, or may become a serious hobbyist, a higher end microscope would be a valuable asset.

Objective lenses: Magnification/Resolution/Field of View/Focal Length

The objective lenses are the most important parts of the microscope. An objective lens not only magnifies the sample, but also determines the resolution. However, higher powered objective lenses with better resolution have a smaller field of view and a shorter focal length.

The resolution and working distance (focal length) of a lens is determined by its numerical aperture (NA). Following is a list of magnifications, numerical aperture, and working distance for some common achromatic objective lenses.

Magnification	Numerical Aperture	Working Distance (mm)
4X	0.10	30.00
10X	0.25	6.10
20X	0.40	2.10
40X	0.65	0.65
60X	0.80	0.30
100X (oil)	1.25	0.18

You can see as the magnification increases the numerical aperture increases (which means the resolution increases) and the working distance decreases.

Choosing the right lens for the right sample is part of the art of microscopy.

Most student projects can be achieved with a 40X objective, however a 100X objective lens can be added to make observing bacteria and small cell structures possible.

Below is a general chart showing the recommended objective lens to use for different types of samples.

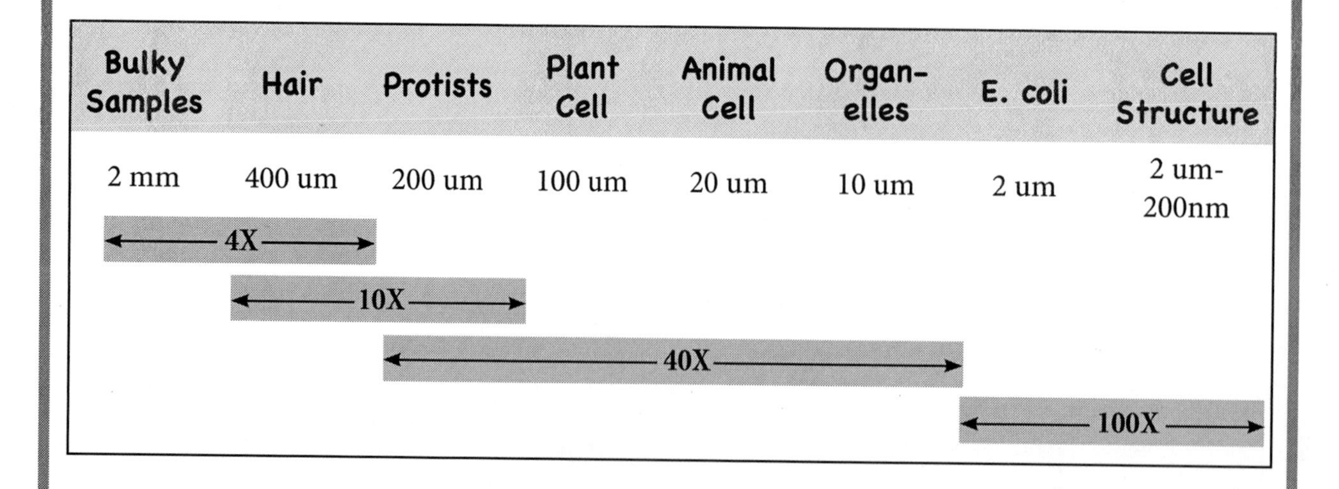

In this experiment students will explore how to use a microscope.

Objectives

The objectives of this lesson are for students to:

- Practice using a microscope.
- Observe small details.

Experiment

I. Think About It

Read this section of the *Laboratory Notebook* with your students.

Ask questions such as the following to guide open inquiry.

- *Do you think a 10X lens will magnify more than a 4X? Why or why not?*
- *Do you think a 100X lens will magnify more than a 10X? Why or why not?*
- *Do you think it will be easier or harder to focus a 4X objective than a 10X objective? Why or why not?*
- *Do you think it will be easier or harder to focus a 10X objective than a 100X objective? Why or why not?*

II. Experiment 7: Using a Light Microscope

Have the students read the entire experiment before writing an objective and a hypothesis.

Objective: Have the students think of an objective for this experiment (What will they be learning?).

Hypothesis: Have the students write a hypothesis. The hypothesis can restate the objective in a statement that can be proved or disproved by their experiment. Some examples:

- *Paper magnified 40X will show fibers.*
- *The ink on paper will look different at 40X and 100X.*
- *I will be able to see blood cells at 40X.*
- *I will only be able to see blood cells at 100X.*

EXPERIMENT

NOTE: **As students are turning the turret to change lenses, help them be extremely careful not to bang the lenses on the stage or glass slide. This can damage the lenses.**

Part I: The Microscope

❶-❽ Have the students follow the instructions in the *Laboratory Notebook* to observe the parts of the microscope and how the different parts work. Have them label the parts of the microscope in the diagram provided.

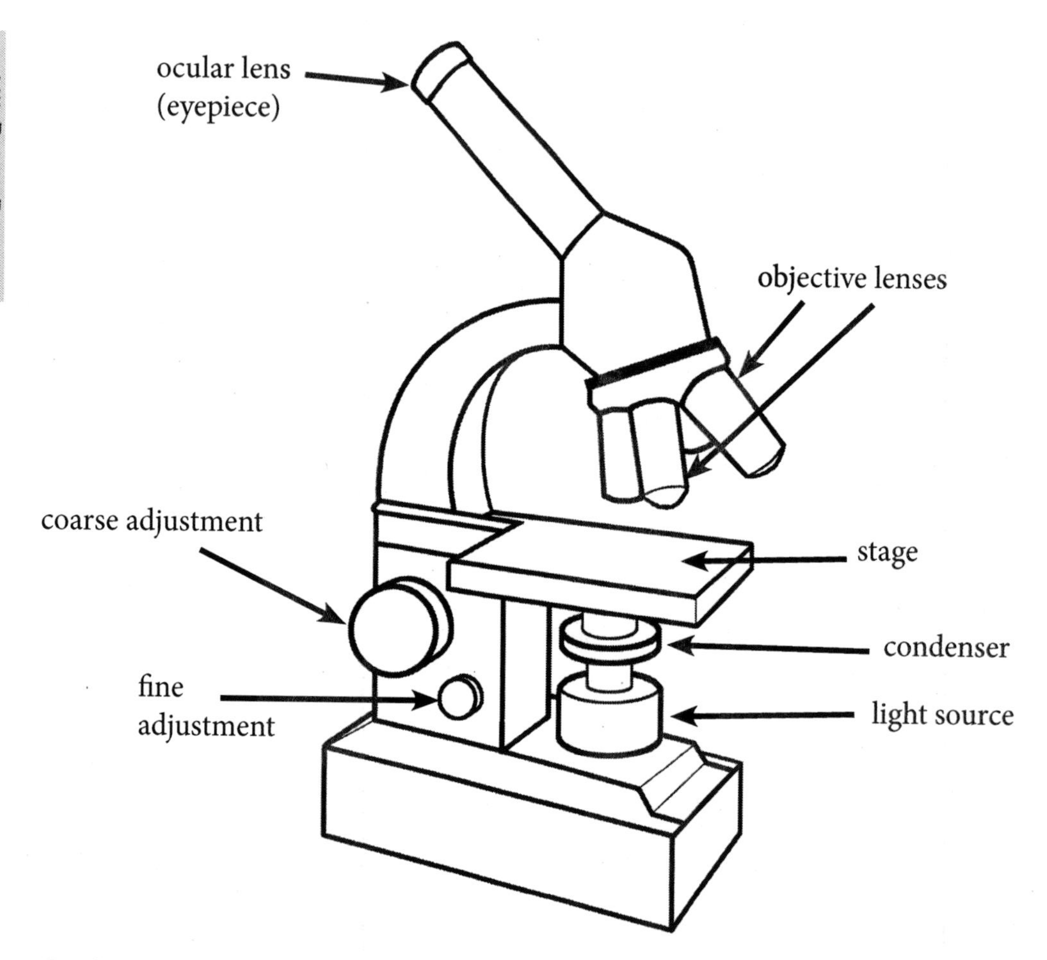

Part II: Observing a Sample

Help the students avoid scraping any of the samples with the objective lenses.

1-6. Have the students take a small piece of paper that has lettering on it and place it on a glass slide in the microscope without a cover slip. Have them examine it using the 4X and 10X objective as instructed. Paper is considered a "bulky" sample, so the low magnification lenses are used. Do not have the students use a higher power lens because it could get damaged. In the spaces provided, have them record their observations.

7. Have the students turn the turret to move the lowest power objective back into place. [NOTE: **If they have a 100X objective, do not let them rotate the turret through this lens. If the lens scrapes the slide, it can ruin the lens.** Instead, have them turn the turret in the opposite direction until the lowest power lens is back in place]

8–14. Have the students create a glass slide with fresh blood. Have the them wash their hands and then collect a drop of blood by pricking a finger with a needle that has been sterilized in a candle flame. Blood is a great sample to observe in a microscope because it will flow for a few minutes before it dries which will coat the area under the coverslip, making it easy to find and focus. Have the students start with the lowest power objective lens (4X) and move to the higher powered objective lenses one at a time, focusing each one as they go. If you plan to have them use a 100X objective lens, help them place a drop of immersion oil on the coverslip and very carefully rotate the turret to click the lens into place. It is very easy to smash the sample surface and ruin the lens, and if this seems about to happen, have them back the lens up one or two turns with the fine adjust knob. Have them always use only the fine adjustment knob with the 100X lens to avoid hitting the slide.

Have them adjust the condenser to get more light.

Each time the students look at the sample through a different lens, have them record their observations in writing and by drawing what they see.

15. Have the students rotate the turret to move the 100X immersion lens away from the sample. Then they can use the coarse adjustment knob to lower the stage and remove the sample.

16. Have them repeat the experiment with other samples, such as hair, pond water, or an insect wing. Have them look at each sample beginning with the lowest power lens and rotating through to the highest power, focusing the image each time before moving to the next highest power lens.

III. Conclusions

Discuss how easy or difficult the students found the use of the microscope. Using a microscope is an art, and learning how to use one correctly takes time and patience. Have the students note whether their conclusions support or do not support their hypothesis.

IV. Why?

Read this section of the *Laboratory Notebook* with your students.
Discuss how the different objectives magnified the samples with different degrees of resolution and focal length.

V. Just For Fun

Have the students use the microscope to observe other samples of their choice. Have them write and draw their observations.

Experiment 8

Observing Protists

Materials Needed

- microscope with a 10X objective
- microscope depression slides [1]
- 10 or more eyedroppers
- fresh pond water or water mixed with soil
- protozoa study kit [2]
- methyl cellulose [3]
- measuring cup and measuring spoons
- baker's yeast
- distilled water
- Eosin Y stain [4]

As of this writing, the following materials are available from Home Science Tools, www.hometrainingtools.com:

1. Glass Depression Slides, MS-SLIDC72 or MS-SLIDC12
2. Basic Protozoa Set, LD-PROBASC (must be used within 1-2 days of arrival)
3. Methyl Cellulose, CH-METHCEL
4. Eosin Y, CH-EOSIN

Objectives

In this experiment students will be introduced to the microscopic organisms known as protists.

The objectives of this lesson are for students to:

- Observe how protists move and eat.
- Use a microscope to make observations.

Experiment

I. Think About It

Read this section of the *Laboratory Notebook* with your students.

Ask questions such as the following to guide open inquiry.

- *How many different ways do you think protists move?*
- *Do you think a paramecium moves more like a euglena or an amoeba? Why?*
- *How many different methods can you think of that protists use to eat?*
- *Do you think using cilia, a flagellum, or pseudopodia is a more efficient way for a protist to move? Why or why not?*
- *Do you think it is easier for a paramecium to get around than for an amoeba? Why or why not?*

II. Experiment 8: Observing Protists—How Do They Move?

In this experiment students will examine the three different types of protists discussed in this chapter of the *Student Textbook*. They will then examine pond water or water mixed with soil to identify individual protists based on their method of movement.

It may take some time for younger students to align their eye directly into the lens so that the sample is visible. Also, viewing tiny organisms through the small eyepiece of a microscope can be difficult and requires some patience. These organisms can swim rapidly through the field of view, and it is easy to get frustrated trying to observe them. Methyl cellulose will help slow the organisms down without killing them. Patience with this experiment is a must.

Because students will be using slides with a concavity for the sample, they will not need to use cover slips.

Have the students read the entire experiment before writing an objective and a hypothesis.

Objective: Have the students write an objective. For example:

- *In this experiment, three types of protists will be observed. We will see how they move in different ways.*

Hypothesis: Have the students write a hypothesis. Some examples:

- *We can tell the difference between ciliates, flagellates, and amoebas.*
- *We can tell the difference between ciliates, flagellates, and amoebas in pond water by how they move.*

EXPERIMENT—Part A

Have the students set up the microscope.

❶ Have the students position a slide in the microscope and use an eyedropper to put a droplet of one of the protist samples on the slide.

❷ Have the students observe how the protists move. A droplet of methyl cellulose can be added to the protist sample on the slide to slow the movement of the protists.

A euglena will tend to move in a single direction, or it may not move at all but "hover" just under the light.

A paramecium will move all over the place. It will roll, move forward and backward, and spin. There are usually other things in the water with the paramecium. Have the students note what happens when the paramecium "bumps" into other objects or other paramecia.

The amoebas move very slowly, and it can be difficult to observe them. They are usually on the bottom of the container. Allow the container to sit for 30 minutes, and then have the students remove some solution from the very bottom, placing it on a slide. The amoebas should be visible but may be difficult to see since they are clear.

❸ Have the students draw the protist they are observing and write their observations. Boxes provided in the *Results* section.

❹ Have the students repeat Steps ❶-❸ using the remaining two protist samples. Have them use a new eyedropper for each sample.

Results—Part A

Boxes are provided for students to record their results.

EXPERIMENT—Part B

Have the students repeat the experiment this time looking for protists in pond water or water mixed with soil. Have them use a new eyedropper to place a droplet of fresh pond water (or water mixed with soil) on a slide in the microscope. Have them look for protists and try to determine the types of protists they observe based on how the organisms move. Have the students refer to the notes they made in *Part A* for comparison. Space is provided in the *Results — Part B* section for writing and drawing their observations. Have them record information for as many different organisms as they can find.

III. Conclusions

Have the students review the results they recorded for the experiment. Have them draw conclusions based on the data they collected. If the experiment did not work, this should be written as a conclusion.

IV. Why?

Read this section of the *Laboratory Notebook* with your students.
Discuss any questions that might come up.

V. Just For Fun: How Do They Eat?

Students will perform another experiment to observe how two different protists eat. Eosin Y stained yeast baker's yeast will be used as food to be ingested by the protists. It may take some time for this observation. Once ingested by a protozoan, the red stained yeast will turn blue.

Have the students read the entire experiment. Have the students predict whether or not the protists will eat the yeast. Then have them write the objective and hypothesis.

❶ Have the students follow the directions to color the yeast with Eosin Y stain:

Add 5 milliliters (one teaspoon) of dried yeast to 120 milliliters (1/2 cup) of distilled water. Allow it to dissolve.

Add one droplet of Eosin Y stain to one droplet of yeast mixture. Look at the mixture under the microscope. You should be able to see individual yeast cells that are stained red.

❷ Have the students place a droplet of the amoeba sample on a glass slide that has been correctly positioned in the microscope. (For the amoebas, remind the students to gather the sample from the bottom of the container it comes in.)

❸ Have the students add a small droplet of the yeast stained with Eosin Y to the droplet of protist solution that is on the slide.

❹ Have the students observe the protists through the microscope, noting the red-colored yeast. Have them describe how the protist eats, writing down as many observations as they can and drawing one of the protists eating. It may take some patience to find protists eating.

❺ Have the students repeat steps 2-4, this time using the paramecium sample.

Experiment 9

Moldy Growth

Materials Needed

- dehydrated agar powder*
- distilled water
- cooking pot
- measuring spoons
- measuring cup
- plastic petri dishes**
- permanent marker
- oven mitt or pot holder
- jar with lid (big enough to hold 235 ml (about 1 cup) liquid
- 1 slice of bread, preferably preservative free
- small clear plastic bag
- white vinegar
- bleach
- borax
- mold or mildew cleaner
- 1-2 pairs rubber gloves

* http://www.hometraining-tools.com/nutrient-agar-8-g-dehydrated/p/CH-AGARN08/

** A stack of 20 can be ordered from: http://www.hometrainingtools.com/petri-dishes-plastic-20-pk/p/BE-PETRI20/

Objectives

In this experiment students will gain more experience in making agar plates and using controls.

The objectives of this lesson are for students to:

- Practice the lab technique of pouring and using agar plates.
- Observe mold growth and products that prevent the growth of mold.

Experiment

I. Think About It

Read this section of the *Laboratory Notebook* with your students.

Ask questions such as the following to guide open inquiry.

- *How do you think mold and bacteria are different?*
- *Do you think products that kill bacteria can kill mold? Why or why not?*
- *Do you think it is easier or harder to prevent mold from growing in wet, warm and dark areas? Why or why not?*
- *Do you think it is easier or harder to prevent mold from growing in dry, cold, and light areas? Why or why not?*

II. Experiment 9: Moldy Growth

Have the students read the entire experiment before writing an objective and a hypothesis.

Objective: Have the students think of an objective for this experiment (What will they be learning?).

Hypothesis: Have the students write a hypothesis. The hypothesis can restate the objective in a statement that can be proved or disproved by their experiment. Some examples include:

- *Bleach will kill mold.*
- *Bleach will not kill mold.*
- *Borax will kill mold better than bleach.*
- *Vinegar will kill mold better than bleach or borax.*

EXPERIMENT

Part I: Pour Agar Plates

Have the students follow the directions in *Experiment* 6 for making agar plates. They will need six plates that don't have bubbles in the agar for this experiment and one or more additional bubble-free plates for the *Just For Fun* section.

Have the students store the plates upside down (agar side up) in the refrigerator until they are ready to do the experiment.

Part II: Observing Mold

In this experiment students will observe which household products kill mold, and whether they prevent growth. The agar on the plates is a food source for the mold from the bread.

❶ Students are to make moldy bread by taking a piece of bread (preferably without preservatives), placing it in a plastic bag, adding a teaspoon of water, sealing the bag, and then letting it sit in a dark, warm area for several days.

❷ Have the students put on rubber gloves. Have them cut 5 small cubes from the moldy bread and save the rest for the *Just For Fun* experiment.

❸-❹ Have the students mark agar plates as follows: *Control, None, Vinegar, Bleach, Borax, and Mildew Cleaner*. They will put the marked plates agar side down.

❺-❻ Students will add 5 ml (1 tsp.) of white vinegar to the *Vinegar* plate and tilt the plate gently back and forth to evenly cover the agar. Have them repeat for the *Bleach* plate.

❼ Have the students measure 235 ml (1 cup) of distilled water into a jar, add 5 ml (1 tsp.) of borax, put the lid on the jar, and shake it until the borax is completely dissolved. They will then measure 5 ml (1 tsp.) of the mixture and put it on the *Borax* plate, tilting the plate back and forth until the agar is covered evenly.

❽ Students will add 5 ml (1 tsp.) of mold cleaner to the *Mold Cleaner* plate and tilt the plate back and forth to evenly cover the agar

❾ Have the students add a cube of moldy bread to each petri dish except the one marked *Control*.

❿ Plates should be moved to a warm, dark room, optimally 27°C (80°F), and stored agar side down.

Results

Have the students observe the plates for several days, and then in the table provided, write and draw their observations.

III. Conclusions

Have the students review the results they recorded for the experiment. Have them draw conclusions based on the data they collected.

Have them compare their results to the control plates (*Control* and *None*) and observe which cleaners killed mold, prevented mold growth, and both killed and prevented mold growth.

IV. Why?

Read this section of the *Laboratory Notebook* with your students.
Discuss how cleaners kill and prevent mold growth. Discuss any questions that might come up.

V. Just For Fun

Part I

Have the students repeat the experiment, this time covering the agar with 5 ml (1 tsp.) of 3% hydrogen peroxide solution and then adding the moldy bread. If they have additional prepared agar plates left, they can try other household products of their choice to see if they kill or prevent mold.

Have them record their results in the space provided.

Part II

Have the students look at a small piece of the moldy bread through their microscope. Have them record their observations.

Experiment 10

Using Electronics

Materials Needed

One of the following recommended electronic circuit kits:

- Snap Circuits: http://www.snapcircuits.net/
  Snap Circuits Jr. 100 Kit
- Little Bits: http://littlebits.cc/intro
  Base Kit: http://littlebits.cc/kits/base-kit

Note: Websites and product availability change over time. If these products are no longer available, do an internet search on children's electronic circuit kits to find a kit suitable for this experiment.

Objectives

In this experiment students will explore electric circuits.

The objectives of this lesson are for students to:

- Learn about basic electric circuits.
- Expand their understanding of electronics in science.

Experiment

I. Think About It

Read this section of the *Laboratory Notebook* with your students.

Ask questions such as the following to guide open inquiry.

- *What do you think your life would be like if there were no electric circuits?*
- *How do you think electric circuits have helped shape the modern world?*
- *How many different items can you name that have an electric circuit?*
- *If you could use electric circuits to create something new, what would you create?*

II. Experiment 10: Using Electronics

Have the students read the entire experiment before writing an objective and a hypothesis.

Objective: Have the students think of an objective for this experiment (what will they be learning?).

Hypothesis: Have the students write a hypothesis. The hypothesis can restate the objective in a statement that can be proved or disproved by their experiment. Some examples:

- *A basic electric circuit can be used to illuminate a light bulb.*
- *A basic electric circuit can be used to rotate a motor.*
- *Electric circuits can be combined*
- *Chemical energy in a battery can be converted to mechanical energy using an electric circuit.*

EXPERIMENT

One of the best ways to learn about electronics is to build electronic circuits. However, putting together electronic circuits can be difficult for most kids, so we recommend buying an electronic circuit kit. Students can then explore the various circuit combinations provided by the kit in a way that is easier for them to understand.

Select one of the electronics kits from the Materials List.

❶-❷ Have your students study the parts and read the instructions and the Do's and Don'ts. Make sure they have a good understanding of what the parts are and how to use them before they proceed with the projects.

❸-❹ Have your students assemble the first two projects in the kit, following the kit instructions. They will be learning more about electric circuits as they work through each project. In the spaces provided, have them record their observations, including a diagram of the finished project with the parts labeled and a description of how the completed project works. Have them note whether the project worked as described.

Making diagrams is an important part of scientific exploration. The diagrams can be sketches rather than realistically drawn since their purpose is to help students understand the projects they build.

Results

Now that the students have a basic understanding of the kit and electric circuits, have them assemble several different projects and record their observations, including labeled diagrams and a description of how each project works. By the time they finish the experiment, they should have a good understanding of the components, how they work, and how to put them together.

III. Conclusions

Have the students answer the questions and draw conclusions based on their observations. Help them determine whether their conclusions support or do not support their hypothesis..

IV. Why?

Read this section of the *Laboratory Notebook* with your students.
Discuss any questions that might come up.

V. Just For Fun

Students are asked to make a circuit of their own design and then record their observations, including a labeled diagram and a description of how it works.

Experiment 11

Moving Marbles

Materials Needed

- several glass marbles of different sizes
- several steel marbles of different sizes
- cardboard tube, .7-1 meter [2.5-3 ft] long
- scissors
- black marking pen
- ruler
- letter scale or other small scale or balance

Objectives

In this experiment students will observe some properties of motion: inertia, friction, and momentum.

The objectives of this lesson are for students to:

- Observe that the motion of an object is changed when an outside force acts on the object.
- Observe how inertia, friction, and momentum together affect the motion of marbles.

Experiment

I. Think About It

Read this section of the *Laboratory Notebook* with your students.

Ask questions such as the following to guide open inquiry.

- *What do you think inertia is?*
- *Do you think you could play baseball without momentum? Why or why not?*
- *Where do you see friction occurring in day to day life?*
- *Do you think friction affects inertia? Why or why not?*
- *Do you think mass affects momentum? Why or why not?*

II. Experiment 11: Moving Marbles

Have the students read the entire experiment before writing an objective and a hypothesis.

Objective: Have the students write an objective. Some examples:

- *We will examine the movement of different marbles.*
- *We will investigate the momentum of different marbles.*
- *We will see what happens when one marble hits another.*
- *We will see if we can move a heavy marble with a light one.*

Hypothesis: Have the students write a hypothesis. Some examples:

- *The small glass marble will not be able to move the steel marble.*
- *The small glass marble will be able to move the steel marble.*
- *The small glass marble will stop when it hits the steel marble.*
- *The small glass marble will not stop when it hits the steel marble.*

EXPERIMENT

❶ Have the students weigh each marble. Have them label each marble with a number or letter or use the color of a marble as identification. Have them record the information for each marble in the chart provided in *Results—Part A.*

Remind the students that weight and mass are different and that they are not going to find the actual mass of the objects. However, they will be able to tell which objects have more mass—that is, those that weigh more. *(See Laboratory Notebook, Section IV. Why?)*

❷ Have the students take the cardboard tube, cut it in half lengthwise to make a trough, measure the length of the trough, and mark the halfway point with the black marking pen.

❸ Have the students measure .3 meter (1 foot) in both directions from the halfway mark and put a mark at each of these measurements. They will then have one mark on each side of the halfway mark.

❹ The cardboard trough should now have three marks: one at the halfway point, and one on either side of the halfway mark, .3 meter (1 foot) away from it. The trough will be used as a track for the marbles.

❺ Have the students roll the marbles one by one down the trough and notice how each one rolls. *(Does it roll straight? Is it easy to push off with your thumb? Does it pass the marks you drew?)* In the space provided in *Results—Part B*, have the students describe how each marble rolls. For example: *The glass marbles move easily down the trough and off the end. The small steel marbles move easily down the trough. The large steel marble takes more effort to get it to move down the trough.* (Answers will vary.)

Students may notice that it takes slightly less effort to push the glass marbles than the heavy steel marbles. This is because larger steel marbles have more inertia than smaller marbles.

❻-❼ Students will next place a glass marble on the center mark of the trough. Have the students roll a glass marble of the same size toward the marble in the center and observe the two marbles as they collide. Have them record their results in *Part C.*

Some example observations are: (Answers will vary.)

- *The rolling glass marble hit the other marble and stopped.*
- *The marble that was stopped started moving when it was hit by the rolling glass marble.*

Ask the students to observe the following:

- *When you roll the glass marble slowly, how far does the marble it hits move?*
- *When you roll the glass marble fast, how far does the marble it hits move?*

What they are observing here is the *conservation of momentum*—the total linear momentum (mass x speed) stays the same. However, because the rolling marble also has angular momentum (converted from linear momentum due to rotation) in addition to linear momentum, and also experiences friction, the stationary marble does not pick up quite all of the total linear momentum of the moving one. In the absence of angular momentum and friction, the total linear momentum of the moving marble would be transferred to the stationary marble. Despite this conversion of some of the linear momentum, the students should still be able to observe qualitatively that, when hit, the stationary marble will move faster when the rolling marble is traveling faster.

❽ Have the students repeat Steps ❻ and ❼ with different size marbles—rolling a heavy marble toward a light marble and a light marble toward a heavy marble.

Help them carefully observe what happens. They should observe that:

- *When the heavy marble impacts the light marble, the light marble will accelerate quite a bit.*
- *When the light marble impacts the heavy marble, the heavy marble will accelerate only a little bit.*

Have the students repeat this several times. Ask them what they think of their results. Have them record their results in *Results—Part D.*

Discuss the *conservation of momentum* which states that the total momentum stays the same. Remind the students that momentum is mass x speed. A large mass traveling at a certain speed will cause a smaller mass to accelerate to a faster speed, and a small mass traveling at a fast speed will accelerate a large mass to a slower speed. Because momentum is conserved, the total momentum will stay the same in each case.

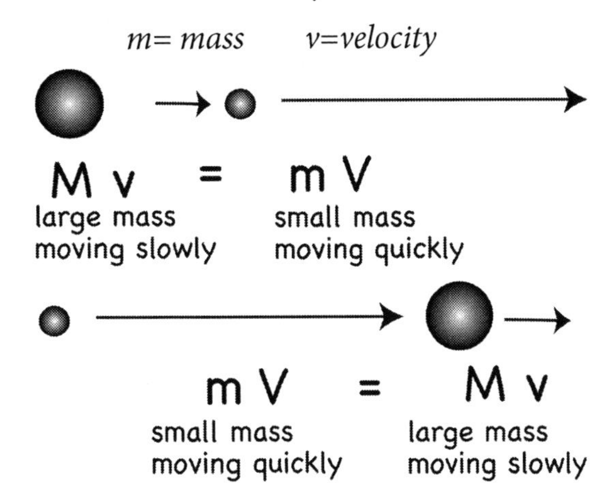

Although there is angular momentum in a rolling ball and friction is also present, the conservation of momentum should again be observable qualitatively in this part of the experiment.

III. Conclusions

Have the students review the results they recorded for the experiment. Have them draw conclusions based on the data they collected.

IV. Why?

Read this section of the *Laboratory Notebook* with your students.
Discuss any questions that might come up.

V. Just For Fun

Students will create their own experiment based on Moving Marbles, this time using a baseball, a basketball, and a golf ball. Have them think about whether they will have to make modifications to the experimental setup. Will they be able to use the cardboard trough? If not, how will they set up the experiment?

Have them record their results and draw conclusions about how differences in mass, inertia, momentum, speed, and friction affect this experiment.

Experiment 12

Accelerate to Win!

Materials Needed

- stopwatch
- compass
- an open space large enough to run (park, schoolyard, playground, backyard, etc.)
- 5 markers of students' choice to mark distances
- blank paper
- a group of friends

Objectives

In this experiment students will explore using basic math in physics formulas.

The objectives of this lesson are for students to:

- Learn how to calculate velocity and acceleration.
- Observe how velocity and acceleration vary over several trials.

Experiment

I. Think About It

Read this section of the *Laboratory Notebook* with your students.

Ask questions such as the following to guide open inquiry.

- *How fast is the fastest human?*
- *How fast is the fastest animal?*
- *How fast do you think you can run?*
- *Do you think you can train to run faster? Why or why not?*

II. Experiment 12: Accelerate to Win!

Have the students read the entire experiment before writing an objective and a hypothesis.

Objective: Have the students think of an objective for this experiment (What will they be learning?).

Hypothesis: Have the students write a hypothesis. The hypothesis can restate the objective in a statement that can be proved or disproved by their experiment. Some examples include:

- *I can run at a steady pace (velocity).*
- *I can accelerate at the end of each trial.*
- *My average velocity will increase with each trial.*

EXPERIMENT

❶ Help the students create a running track that is straight and has a manageable length. Have them mark a starting point and an ending point for the distance they will run. Make sure that when the total length of the track is divided into fourths, there will be enough time during the run for a timer to record the time of each segment of the run (one-fourth of the total track length).

❷ Have the students use a compass to determine in which direction they will be running and then record this direction in the chart in the *Results* section.

❸ Students will measure the length of the track using their own feet, walking heel-to-toe from the starting point to the finish with each step being one "foot." In the space provided, have them record the distance they measure.

❹ Using their measurement of the total distance to be run, students will calculate the distance of one-fourth of the track and then record this result for each of the time points $\mathbf{d_1}$-$\mathbf{d_4}$. This is the distance they will run for each of the four segments of the track.

❺ Have the students use their feet to measure the distance between the time points and mark each time point on the track in a way that the timer can see the mark as the runner is passing it.

❻-❼ One person will use a stopwatch to clock the times. Have them stand in a position where they will be able to see all the time points as the runner passes them. A second person will record the times in the *Laboratory Notebook*. Have your students run from the starting point to the finish line with the time keeper recording the time at each time point.

❽ Have the students run three or four times or until they are too tired to continue.

Results

❶-❷ Have your students calculate the velocity of each segment for each time trial and the acceleration between time points for each trial. Formulas are provided in the chart where students will record their answers. An example of calculations is included on the following page in this *Teacher's Manual*.

Space is provided for doing the calculations, but extra paper may be needed. If students use extra paper, have them fasten it in the *Laboratory Notebook* when they have completed the experiment.

A Place for Calculations (Calculations for example on next page)

Velocity

Trial 1: v_1 $v_1 = \frac{d_1}{t_1} = \frac{150 \text{ ft.}}{11 \text{ sec.}} = 13.6 \text{ ft./sec.}$

Trial 1: v_2 $v_2 = \frac{d_2}{t_2} = \frac{150 \text{ ft.}}{12 \text{ sec.}} = 12.5 \text{ ft./sec.}$

Trial 1: v_3 $v_3 = \frac{d_3}{t_3} = \frac{150 \text{ ft.}}{11 \text{ sec.}} = 13.6 \text{ ft./sec.}$

Trial 1: v_4 $v_4 = \frac{d_4}{t_4} = \frac{150 \text{ ft.}}{9 \text{ sec.}} = 16.7 \text{ ft./sec.}$

Acceleration*

Trial 1: $a_1 = \left(\frac{v_2 - v_1}{|t_2 - t_1|}\right) = \frac{12.5-13.6}{|12-11|} = \frac{-1.1 \text{ ft./sec.}}{1 \text{ sec.}} = -1.1 \text{ ft. sec.}^2$ (squared)

Trial 1: $a_2 = \left(\frac{v_3 - v_2}{|t_3 - t_2|}\right) = \frac{13.6-12.5}{|11-12|} = \frac{1.1 \text{ ft./sec.}}{1 \text{ sec.}} = 1.1 \text{ ft. sec.}^2$ (squared)

Trial 1: $a_3 = \left(\frac{v_4 - v_3}{|t_4 - t_3|}\right) = \frac{16.7-13.6}{|9-11|} = \frac{3.1 \text{ ft./sec.}}{2 \text{ sec.}} = 1.55 \text{ ft. sec.}^2$ (squared)

***Note:** For acceleration, time is always a positive number. In the acceleration formula, the change in time (Δt) is written as $|t_f - t_i|$ to show that the result is expressed as a positive number.

PHYSICS

Time Trial Results (Example—answers will vary)

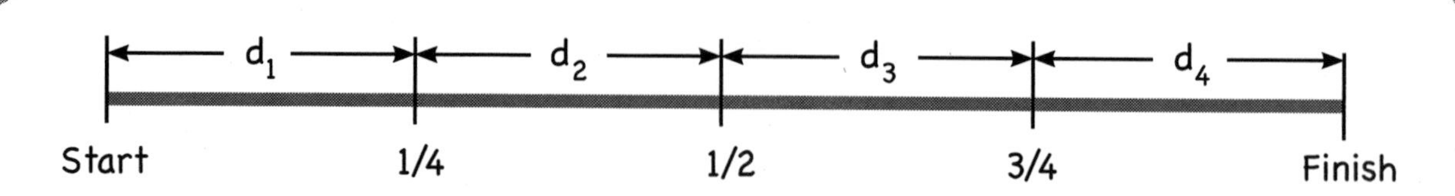

Direction *north*

Distance (in "feet")

d_1	d_2	d_3	d_4
150	*150*	*150*	*150*

Time (seconds)

	t_1	t_2	t_3	t_4
Trial 1	*11*	*12*	*11*	*9*
Trial 2				
Trial 3				

Velocity

	$v_1 = \frac{d_1}{t_1}$	$v_2 = \frac{d_2}{t_2}$	$v_3 = \frac{d_3}{t_3}$	$v_4 = \frac{d_4}{t_4}$
Trial 1	*13.6 ft./sec.*	*12.5 ft./sec.*	*13.6 ft./sec.*	*16.7 ft./sec.*
Trial 2				
Trial 3				

Acceleration*

	$a_1 = \left(\frac{v_2 - v_1}{\lvert t_2 - t_1 \rvert}\right)$	$a_2 = \left(\frac{v_3 - v_2}{\lvert t_3 - t_2 \rvert}\right)$	$a_3 = \left(\frac{v_4 - v_3}{\lvert t_4 - t_3 \rvert}\right)$
Trial 1	*-1.1 ft. sec.^{2}*	*1.1 ft. sec.^{2}*	*1.55 ft. sec.^{2}*
Trial 2			
Trial 3			

***Note:** For acceleration, time is always a positive number. In the acceleration formula, the change in time (Δt) is written as $\lvert t_f - t_i \rvert$ to show that the result is expressed as a positive number.

III. Conclusions

Have the students review the results they recorded for the experiment, answer the questions, and then draw conclusions based on their observations. Have them note if their conclusion supports or does not support their hypothesis.

IV. Why?

Read this section of the *Laboratory Notebook* with your students.
Discuss any questions that might come up.

Discuss with your students how they can measure their own velocity and acceleration by using time and distance. Explain how knowing these numbers can help a coach train an athlete for the Olympics. Have them discuss how this information might help them in a sport they participate in or how it might be used by participants in a sport they're interested in.

V. Just For Fun

Have the students run every day and record the date and amount of time they spend on each run. A chart is provided. They may find it interesting to create their own chart on a separate piece of paper that has a schedule for the runs and space to record the date and length of time of each run along with observations including how difficult or easy each run was, their route and its characteristics, the weather, etc. Do they notice a difference between the first and last training runs? What factors affect their runs?

After a few weeks have them repeat the experiment. Have them record and calculate their results and compare them to the first set of trials. What differences in the trials can they notice? Do they think running every day for a few weeks made a difference in the results? Has using the physics in this experiment helped them better understand how to run a race? Has it helped them observe a difference in their fitness? What conclusions can they draw?

Experiment 13

Around and Around

Materials Needed

- pencil or pen
- marking pen
- thumbtack or pushpin
- 3 pieces of string—
  approximate sizes:
  10 cm [4 in.]
  15 cm [6 in.]
  20 cm [8 in.]
- tape
- ruler (metric)
- large piece of white paper (bigger than 30 cm [12 in.] square—students may need to tape several sheets of paper together)
- firm surface at least as large as the paper and that a thumbtack can be pinned into

Objectives

In this experiment students gain a better understanding of tangential speed and how it can be calculated mathematically.

The objectives of this lesson are for students to:

- Explore how math can be used to understand physics.
- Use math to calculate tangential speed.

Experiment

I. Think About It

Read this section of the *Laboratory Notebook* with your students.

Ask questions such as the following to guide open inquiry.

- *What does it feel like to ride on a playground merry-go-round? Why?*
- *As the merry-go-round is spinning, does it feel different if you sit near the center than if you sit near the outer edge? Why or why not?*
- *If you run in a circle around your room, how far do you go?*
- *If you run in a circle around your house, how far do you go? Is it farther than when you run around your room?*
- *How far do you think you would go if you could run around the Earth? the Sun? the solar system?*
- *If you could run around your house, the Earth, and the Sun, what speed do you think you would have to go to circle each in the same amount of time? Why?*

II. Experiment 13: Around and Around

Have the students read the entire experiment before writing an objective and a hypothesis.

Objective: Have the students think of an objective for this experiment (What will they be learning?).

Hypothesis: Have the students think about what they might learn from calculating the tangential speed of a circular path.

EXPERIMENT

❶ Have the students measure and cut the three lengths of string and, if necessary, tape together several sheets of white paper to a size larger than 30 cm [12 in.] square.

Provide a firm, flat surface at least as large as the paper and that a thumbtack or pushpin can be pinned into. Have the students fasten one end of the shortest string to the thumbtack or pushpin. Help them think about how this can be done; for example, making a loop in the string, making a knot at the end of the string, or wrapping the string around the thumbtack.

❷ Have the students use the marking pen to put a mark on the string at 5 cm (2 in.) from the thumbtack.

❸ Have the students place the pen or pencil at the 5 cm (2 in.) mark on the string, wrap the extra string around the pen and fasten it with tape.

❹- ❺ Have the students hold the thumbtack down so it doesn't dislodge and pull the pen away from the thumbtack until the string is stretched out to its full length. The pen should be held perpendicular as it is placed on the paper surface. Then have the students draw a circle with the thumbtack at the center. This will result in a 5 cm (2 in.) radius for the smallest circle.

❻ Have the students repeat Steps ❶-❺ with the other two pieces of string. For the middle size string the pen will be 10 cm (4 in.) from the thumbtack, and for the longest piece of string the pen will be 15 cm (6 in.) from the thumbtack.

Results

A table is provided with formulas for calculating the circumference of the circles and the tangential speed (see next page).

❶ Have the students measure the radius of each circle. Point out that the radius is the same as the length of the string. Have them record the radii in the table.

❷ Have the students calculate the circumference. They are provided with the equation: $c=2\Pi r$.

❸ Have the students calculate the tangential speeds using the circumference of each circle and assuming one rotation takes 1 minute (1 RPM).

Calculating Tangential Speed

	①	②	③
String Length	5 cm	10 cm	15 cm
Circle:			
Radius	*5 cm*	*10 cm*	*15 cm*

Calculate the Circumference [Π (pi) = 3.14]

①	②	③
c = 2Πr	**c = 2Πr**	**c = 2Πr**
c = 2Πr · *5 cm*	c = 2Πr · *10 cm*	c = 2Πr · *15 cm*
c = *31.4 cm*	c = *62.8 cm*	c = *94.2 cm*

	①	②	③
Circumference	*31.4 cm*	*62.8 cm*	*94.2 cm*

Calculate the tangential speed for one revolution (1 RPM)

Tangential speed = distance traveled/time

Note: Distance traveled is one revolution—the circumference of the circle (c).

Time (t) equals one minute for this problem.

Tangential speed (S_T) = c/t

Calculation

①	②	③
S_T = c/t	**S_T = c/t**	**S_T = c/t**
S_T = *31.4* /t	S_T = *62.8* /t	S_T = *94.2* /t

	①	②	③
Tangential Speed	*31.4 cm/min.*	*62.8 cm/min.*	*94.2 cm/min.*

III. Conclusions

Have the students answer the questions and draw a conclusion based on their observations. Note if their conclusion supports or does not support their hypothesis.

IV. Why?

Read this section of the *Laboratory Notebook* with your students.

Discuss with your student how using mathematics helps us explain many of the phenomena we experience in the real world, such as spinning on a merry-go-round.

Discuss any questions that might come up.

V. Just For Fun

Have your students think of some different objects that show tangential speed when in motion and list these objects in the space provided.

Next, have them think of a way to calculate the tangential speed of one of these objects. For example, a tape measure can be used to measure the radius at different points from the center of a merry-go-round. The radii can then be plugged into the equation provided.

Experiment 14

Hidden Treasure

Materials Needed

- pencil, pen, colored pencils
- compass
- a small jar or container with a lid
- small items to place in jar (student selected treasure)
- garden trowel (optional)

Objectives

In this experiment students will explore some tools geologists use to study Earth.

The objectives of this lesson are for students to:

- Explore mapmaking.
- Use a compass to find directions.

Experiment

I. Think About It

Read this section of the *Laboratory Notebook* with your students.

Ask questions such as the following to guide open inquiry.

- *What different kinds of maps have you used? What did you use them for? What else do you think maps could be used for?*
- *What parts of making a map do you think would be easy? Difficult? Why?*
- *Why do you think geologists need to use maps?*
- *What problems could occur if you were using a map that was not accurate? Why?*
- *When do you think it would be helpful to use a compass? A GPS? Why?*
- *What advantages do modern geological tools have over older tools?*
- *What disadvantages might modern tools have?*
  (For example, the need for electricity or gas power, fragility, need for software updates, expense, etc.)

II. Experiment 14: Hidden Treasure

In this experiment students will explore mapmaking by creating a map for finding a hidden treasure. Students will then test the accuracy of their map by having a friend try to find the treasure that is hidden or buried.

Have the students read the entire experiment before writing an objective and a hypothesis.

Objective: Have the students write an objective. Some examples:

- *To explore mapmaking by creating a real map.*
- *To understand how maps are made and the difficulties that come up.*
- *To create and test a homemade map.*

Hypothesis: Have the students write a hypothesis. Some examples:

- *I can create a map that is 80-100% accurate and I will know this by how quickly my friend finds the treasure.*
- *By creating a map and having it tested, I will learn about the difficulties of mapmaking.*
- *I can test the accuracy of my map by having a friend find my buried treasure.*

EXPERIMENT

In this experiment students will make a map and use a compass to determine north, south, east, and west.

❶ Have the students gather some small objects to use as treasure and place them in a jar.

❷ The students will be making their own map to use in finding a treasure they will hide. Help the students select an area to map—the backyard, a park, or other open space.

❸-❹ Students are to draw the outline of their map in the space provided. They will be using their footsteps as a tool for making measurements to put on their map. Have them count their steps while walking heel-to-toe around a given space. This space can be square or rectangular or oddly shaped. Have them draw on the map the actual shape of each side of the space and note the measurements and any obstacles they may encounter as they measure the space.

❺ Help the students understand how to use a compass.

Using a Compass

The needle on a compass will always point north (N). To read a compass, line the needle up with the north (N) label. If you wanted to travel north, you would head in this direction. To go south, you would travel in the opposite direction.

Once you have determined which way is north, you can find any other direction in which you want to go. For instance, if you want to go northeast, you would line the compass needle up with the north label and then travel in the direction of the northeast (NE) label on the compass.

Indicating Directions on the Map

Students will note the four directions on their map, indicating north with "N" along with an arrow. Then they will mark south with an "S", east "E," and west "W," again drawing arrows.

It may be helpful to the students to have them place their map on the ground in the same orientation as the landscape features. Have them hold the compass and stand next to the map. Direct them to turn until the compass needle points to north (N) and then draw an arrow and an N on their map to indicate which way is north. North will probably not be at the top of their map.

Explain that south is opposite north and have them mark south on their map with an arrow and an S. Next explain that when facing north, east will be to the right and west to the left. Have them mark east and west on their map.

You can also have the students use the compass to determine the direction each side of their map is facing.

❻ Have the students add details to their map, indicating objects and distances between them.

❼ Have the students pick a location to bury or hide their treasure and indicate this on their map.

❽ Students are to give their map to a friend and see if the friend can use it to find the treasure.

Results

Have the students record the number of attempts their friend makes in order to locate the buried treasure. Also have the students note any help they have to give to the friend in finding the treasure. This help can be used to make adjustments to the map to make it more accurate.

For example:

- *The number of steps was counted or recorded incorrectly.*
- *Direction of travel was drawn or followed incorrectly.*
- *An object noted on the map was moved.*

III. Conclusions

Have the students review the results they recorded for the experiment. Have them draw conclusions based on the data they collected.

IV. Why?

Read this section of the *Laboratory Notebook* with your students.
Discuss any questions that might come up.

V. Just For Fun

Have the students evaluate their map. Ask questions to help with their evaluation. For example:

- *Was the map accurate enough for your friend to find the hidden treasure? Why or why not?*
- *What (if any) modifications to the map did you make? Can you think of other ways to improve your map?*
- *How might you make your map more accurate?*
- *Does the size of your feet compared with the size of your friend's feet make a difference?*

Have the students review the results of their experiment. Encourage them to think of ways the map could be improved and then revise the map, including a new location for the hidden treasure. Have them give the map to a friend and see how well the revised map works compared to the original map.

Have the students evaluate how well the revised map worked compared to the original version. Have them record their results in the space provided.

Experiment 15

Using Satellite Images

Materials Needed

- computer with internet access (a program that unzips files may be needed)

Optional

- printer and paper
- colored pencils

Objectives

In this experiment students will explore satellite images and how they are used by geologists.

The objectives of this lesson are for students to:

- Learn how to use online resources to collect data.
- Observe how technology such as satellite imagery can be used to study changes in Earth's system.

Experiment

I. Think About It

Read this section of the *Laboratory Notebook* with your students.

Ask questions such as the following to guide open inquiry.

- *What do you think might happen to a beach during a hurricane? Which of Earth's spheres do you think could be affected by a hurricane?*
- *What do you think might happen to trees during a forest fire? Which of Earth's spheres do you think could be affected by a forest fire?*
- *What do you think might happen to rivers and lakes during a drought? Which of Earth's spheres do you think could be affected by a drought?*
- *How many of Earth's spheres do you think can be affected when a volcano erupts? Which ones? How?*

II. Experiment 15: Using Satellite Images

Have the students read the entire experiment.

Objective: Have the students think of an objective for this experiment (What will they be learning?).

Hypothesis: There is no hypothesis for this experiment since it is an observational experiment.

EXPERIMENT

❶ Have the students go to the US Geological Survey (USGS) website and spend some time looking through the collections of sets of satellite images that show changes to Earth's surface over time. As of this writing, the URL for the images is:

https://remotesensing.usgs.gov/gallery/

❷-❺ Students will be looking for images that they think show changes that have affected each of the different spheres of Earth. Have them select one set of images for each of the spheres: the geosphere, the atmosphere, the hydrosphere, and the biosphere.

Under each set of images on the USGS website there will be a *Download Image* section. Have the students click on the small size image file and download it. If they get a "zip file" that contains the images, they will need a program that unzips files for them to be able to look at the images. Have them make a folder on the computer and put the downloaded image files in it.

If possible, have them print the images, label them, and insert them in their *Laboratory Notebook*. If a printer isn't available or the images don't print well enough to be usable, students can refer to the files on the computer as needed.

Observations will be noted in the *Results* section.

Results

For each sphere, have the students fill in the information requested in the charts. They are asked to make a rough sketch of the images they are referencing even though they may have printed them out. Making sketches will help the students look more carefully at the images. The sketches do not need to be highly detailed and can be done in color or black and white. They can be done on separate pieces of paper if students want a larger area for drawing.

Note: Websites change from time to time. Should the referenced USGS website become unavailable, check to see if the Landsat images are now in a different location on the USGS site. If you can't find them, students can use one of the websites in the *Just For Fun* section. In this case, they probably won't be seeing side-by-side before and after images, but they should still be able to find images where changes have occurred due to storms, volcanic eruptions, etc.

Another possible resource for satellite imagery is the European Space Agency (ESA) site: http://www.esa.int/Our_Activities/Observing_the_Earth/

III. Conclusions

Have the students discuss what they learned by observing images of Earth taken from space and how they think satellites have changed what we know and can discover about Earth.

IV. Why?

Read this section of the *Laboratory Notebook* with your students. Discuss with your students how satellite images can help scientists observe Earth's interconnected systems and how natural and human activities interact with these systems. Discuss any questions that might come up.

V. Just For Fun

Students will spend some time viewing more satellite images on two NASA websites—or you can have them choose one of the sites. Have the students follow the directions in the *Laboratory Notebook* for using each website. The Earth Observatory website is the more straightforward of the two.

NASA's Earth Observatory website has photographs and videos taken from the International Space Station.

http://earthobservatory.nasa.gov/

Directions provided to students: On the top menu bar click on *Images* to find collections by topic. Select *Natural Hazards* and explore satellite images of several natural disasters. Go back to the Home pages and scroll down to find the *Special Collections* groups of photographs to explore.

NASA's Gateway to Astronaut Photography of Earth

http://eol.jsc.nasa.gov/Collections/EarthFromSpace/

Directions provided to students: Select a topic from the left menu bar. On the next screen select what you would like to view and click *Start Search*. Play around to see what else you can find on this site.

Have the students answer the questions about their observations. Have them discuss the new things they discovered.

Note: Websites change from time to time. If the above referenced websites work differently, help the students navigate to the information they are looking for. Should the above referenced websites become unavailable, browser searches can be done to find other NASA or ESA websites that have satellite images of Earth.

Experiment 16

Modeling Earth's Layers

Materials Needed

Some suggestions for student chosen model making materials:

- modeling clay of different colors
- marble or steel ball
- ingredients to make various colored cakes
- materials for making paper mache
- Styrofoam balls

Objectives

In this experiment students will explore model making to learn more about the layers of the geosphere.

The objectives of this lesson are for students to:

- Explore the features of the geosphere by making a model.
- Observe the uses and limitations of model making.

Experiment

I. Think About It

Read this section of the *Laboratory Notebook* with your students.

Ask questions such as the following to guide open inquiry.

- *What types of models have you built?*
- *If you could build a model of anything you'd like to, what would you build?*
- *How easy is it to make a model of an airplane, car, bicycle, or skateboard?*
- *Why do you think people build models?*
- *How has model building helped you better understand the object you are modeling?*

II. Experiment 16: Modeling Earth's Layers

Have the students read the entire experiment.

Objective: Have the students think of an objective for this experiment (What will they be learning?).

Hypothesis: Have the students think of a hypothesis for this experiment.

EXPERIMENT

1. Have the students list the different layers and the features associated with each layer (e.g., outer crust—hard, made of rock; lithosphere—brittle, divided into plates, etc.)

2. Brainstorm with your students to help them think of different ways to create a model of Earth and Earth's layers. Have them review their chart and select features they want to model and help them select appropriate materials. For example, if they want to explore the spherical nature of Earth's layers, they can use modeling clay, paper mache, Styrofoam balls, etc. If they want to explore the different textures of Earth's layers, they might ignore the spherical aspect of Earth and make their layers in a flat pan using gelatin, peanut butter, cake layers, etc.

3. Have the students use their ideas to build a model of the layers of the geosphere.

Results

1. Have the students draw a diagram of their model, label the parts, and list the materials used to make each layer.

2. Help the students think about which features they decided to include and which they decided to leave out and why.

III. Conclusions

Have the students think about what they learned by building a model of Earth's layers and discuss the limitations of the model they constructed.

IV. Why?

Read this section of the *Laboratory Notebook* with your students.

Discuss with your student how model building is important for science but that models do not give exact examples of the ideas and objects that are being modeled.

V. Just For Fun

Help the students think about what it would take to model S or P waves moving through Earth's layers. Have them come up with a flat model and have a wave pass through it. Encourage them to use their imagination in coming up with materials. For example, they might lay a small throw rug on the floor and shake one end to get a wave to pass through it. They might also place items on top or under it to see how they are affected.

Experiment 17

Exploring Cloud Formation

Materials Needed

- 2 liter (2 quart) plastic bottle with cap
- warm water
- matches
- blank paper

Objectives

In this experiment students will explore how clouds are formed.

The objectives of this lesson are for students to:

- Understand that different factors in the atmosphere affect cloud formation.
- Use simple tools for exploration.

Experiment

I. Think About It

Read this section of the *Laboratory Notebook* with your students.

Ask questions such as the following to guide open inquiry.

- *What do you think Earth would be like if it did not have an atmosphere?*
- *Do you think life could survive on Earth if the atmosphere did not carry water? Why or why not?*
- *Do you think the atmosphere is involved in the creation of rain and snow? Why or why not?*
- *Why do you think atmospheric pressure is important to life?*
- *What do you think would happen to the atmosphere if Earth's gravity were weaker? Would this affect clouds? Why?*

II. Experiment 17: Exploring Cloud Formation

Have the students read the entire experiment.

Objective: Have the students write an objective. Some examples:

- *To make clouds in a bottle.*
- *To understand how clouds are made.*
- *To see what happens to air and water in a bottle if a lit match is thrown in.*

Hypothesis: Have the students write a hypothesis. Some examples:

- *The air in the bottle will make a cloud.*
- *The air in the bottle will change as the bottle is squeezed and released.*
- *The lit match will go out when it hits the water.*

EXPERIMENT

1. Have the students pour warm water into the plastic bottle until it is about 1/4 full and then put the cap on the bottle. Putting the cap on the bottle will cause water vapor to form.
2. Have the students light a match, remove the cap from the bottle, and then drop the match in the bottle and quickly replace the cap. Students may need help with this step since the bottle cap must be removed and replaced quickly in order to keep water vapor from escaping.
3. Have the students squeeze the plastic bottle near the bottom and release while observing what happens to the air in the bottle. Squeezing the bottle will increase the air pressure inside. Releasing the bottle will cause the air pressure to decrease, lowering the air pressure and thus cooling the air as it expands. This should cause clouds to appear. If clouds don't appear, have the students try using warmer water.
4. Have the students record their observations in the chart in the *Results* section. Have them write down as many details as possible.
5. Have the students repeat this experiment, filling the plastic bottle 1/2 full, 2/3 full, and then almost full, emptying the bottle each time and starting with fresh warm water. Have them record their observations each time.

Results

A chart is provided for the students' observations.

III. Conclusions

Have the students review the results they recorded for the experiment. Have them draw conclusions based on the data they collected.

Based on the observations they have recorded, have the students write down whether they think any of the variables might have changed during the different steps of the experiment. Explain that *ratio* means the quantity of the different factors relative to each other.

Ask questions such as the following:

- *Was the water the same temperature for each part of the experiment?*
- *Was the cap on or off for a longer time?*
- *Was the bottle harder or easier to squeeze?*
- *Was there more or less air in the bottle?*
- *Was there more or less smoke?*

IV. Why?

Read this section of the *Laboratory Notebook* with your students.
Discuss any questions that might come up.

V. Just For Fun

Students are asked to observe the clouds daily for two weeks or more and record their observations of the clouds and the weather conditions. Have them include sketches of the clouds they see. A chart format is provided for them to follow as they make additional chart pages on blank paper.

Have them check online for the humidity, dew point, and low and high temperatures and record this data daily on their chart.

At the end of the experiment, have the students analyze the data they have recorded to look for relationships between weather conditions, types of clouds, temperature, humidity, and dew point. Have them record their conclusions and fasten their chart pages in their *Laboratory Notebook*.

Experiment 18

Measuring Distances

Materials Needed

- two sticks (used for marking locations)
- two rulers
- tape
- string, several meters long (several yards)
- protractor

Objectives

This experiment introduces students to the concept that tools and math can be used to measure the distance to faraway objects.

The objectives of this lesson are to have students:

- Understand that one of the most important elements of scientific investigation is the use of proper tools for studying the world around us.
- Understand that math concepts have helped shape our understanding of the universe.

Experiment

I. Think About It

Read this section of the *Laboratory Notebook* with your students.

Ask questions such as the following to guide open inquiry.

- *Do you think you can tell just by looking at the night sky which stars are closest and which are farthest away? Why or why not?*
- *Do you think astronomers need to use tools to measure the distance to a star? Why or why not?*
- *What kinds of tools do you think astronomers use to study objects in space? What can they discover by using these tools?*
- *When do you think it would be most important for an astronomer to know how far it is to the Moon or a planet? Why?*
- *What do you think can be learned about celestial bodies by looking through a telescope? Why?*

II. Experiment 18: Measuring Distances

Have the students read the entire experiment before writing an objective and a hypothesis.

Objective: Have the students write an objective. For example:

- *To determine the distance of a faraway object using the method of triangulation.*

Hypothesis: Have the students write a hypothesis. For example:

- *The distance of an object can be found using triangulation.*

EXPERIMENT

❶ This experiment is done in an open space such as a field, city street, or backyard. A distant object needs to be visible. Students will be measuring the distance to this object using triangulation. The open space needs to be wide enough for students to locate two observation points some distance from each other with each point having a line of sight to the distant object.

❷ Have the students choose two points from which they can see the distant object. Have them mark each observation point with a stick and mark one "A" and the other "B." If the students are on dirt or grass, the sticks can be put in the ground. Rocks or other markers can be used in place of sticks.

❸ Have the students tape two rulers together at one end, making a right angle.

❹ Have the students place the corner of the taped rulers on observation point "A" with one end pointing towards the distant object, the other end pointing towards observation point "B."

❺ Have the students attach the string to the stick at observation point "A," and stretch it out along the side of the taped ruler that is pointing towards observation point "B." The string will help the students walk in a straight line.

❻ Have the students hold the string and walk heel-to-toe from observation point "A" to observation point "B" while counting their steps. Each of their feet is used as a measurement device (one "foot"). Help the students keep the string pointing at a 90° angle while they are walking in the direction of point "B." (Note: Students may need to adjust the location of point "B" to maintain the right angle at point "A.")

❼ When the students get to point "B," have them attach the string to the stick and make sure the string is still pointing in the same direction as the ruler to maintain the 90° angle. In the space provided in the *Results* section, have them record the number of steps between point "A" and point "B."

❽ Have the students place the protractor on the string. Have them look toward the distant object and observe the angle of view indicated on the protractor.

❾ In the *Results* section, have the students record the angle between point "B" and the distant object.

Experimental Setup

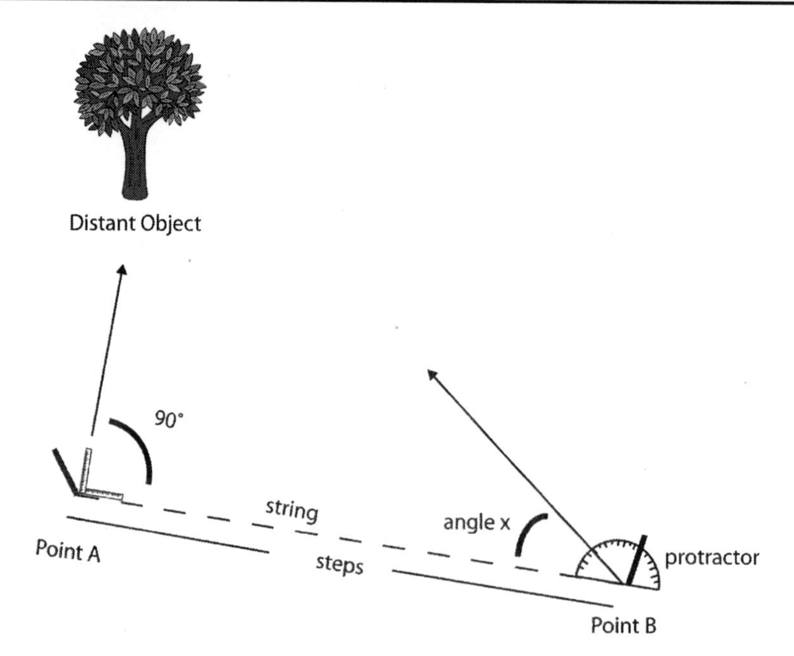

Results

Grid paper is provided. Students will use the data they collected and the triangulation technique to measure the distance to the object. See example graph on next page.

❶-❷ Have the students mark point "A" on the grid paper, count one square for each step they took, and then mark point "B." If the grid paper isn't wide enough, they can assign a different unit to each square; for instance, one square equals 2 feet.

❸ Have the students draw a line connecting the two points. This is line AB.

❹ Have the students draw a line from point "A" towards the distant object at a 90° angle to line AB and label this line "y."

❺ Have the students use the protractor to mark the angle they measured at point "B" to the distant object. Have them draw a line from point "B" until it intersects with line "y." They may have to extend line "y" in order for the two lines to intersect. The intersection point is the location of the distant object.

❻-❼ Have the students count the number of squares from point "A" along line "y" to the distant object. Assuming that each of their steps is one foot, this will be the distance to the object. Have the students record their answers.

Example (Answers will vary.)

Number of steps — Point A to Point B *19 steps*

Angle at Point B *45 degrees*

Number of squares — Point A to distant object *13 squares*

Distance of object in feet *13 feet*

Example:

The graph should look something like the one below, although the actual distances and angles may vary. Make sure that point A has a 90° angle between line **y** and line **AB**. The students will use the protractor to measure angle **B**. Make sure the angle recorded on the graph is the same as the angle measured in Step ❽ of the *Experiment* section. Students may need to adjust the length of line **y**. Point out to the students that for smaller angles, line **y** will be shorter and for larger angles, line **y** will be longer.

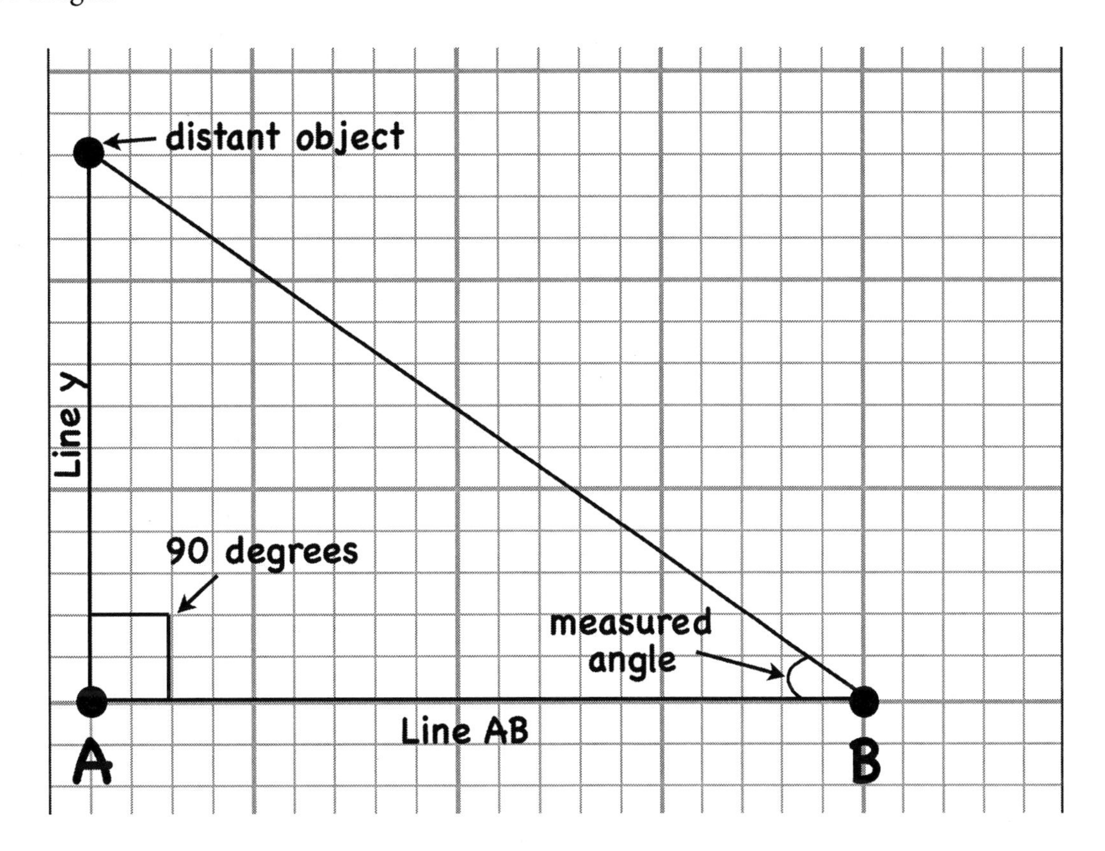

III. Conclusions

Have the students review the results they recorded for the experiment and discuss possible conclusions. Encourage discussion with questions such as the following:

- *How accurate do you think your measurement is?*
- *Could you use a different method to confirm your results? If so, what would you do?*
- *What sources of error do you think might affect your result?*

IV. Why?

Read this section of the *Laboratory Notebook* with your students.
Discuss any questions that might come up.

V. Just For Fun

Part A. Students will choose a different distant object and repeat the experiment. Help them find a distant object that has an unobstructed path to it. Have them record their data.

Part B. This time they will be verifying their calculated distance to the object by counting their heel-to-toe steps while walking from point A to the object. Have them record and analyze their results.

Experiment 19

Using a Star Map

Materials Needed

- computer with internet access
- printer and paper
- flashlight

Optional

- binoculars or telescope
- star map app and mobile device

Objectives

In this experiment students will learn how to use a star map to identify stars.

The objectives of this lesson are to have students:

- Learn how to use a star map.
- Observe how a star map changes with the months and seasons.

Experiment

I. Think About It

Read this section of the *Laboratory Notebook* with your students.

Ask questions such as the following to guide open inquiry.

- *How many stars do you think are in the night sky? Why?*
- *Do you think you could see more stars with a telescope than with your unaided eyes? Why or why not?*
- *If your job was to map all the stars you could see, how would you do it? Do you think there is more than one way to map stars?*
- *Do you think astronauts need to know anything about the stars? Why or why not?*

II. Experiment 19: Using a Star Map

Have the students read the entire experiment before writing an objective and a hypothesis.

Objective: Have the students think of an objective for this experiment (What will they be learning?)

Hypothesis: This is an observational experiment so there is no hypothesis.

EXPERIMENT

❶ Have your students go to the Starmap site at http://www.star-map.fr/ and click on the *Free Maps* menu tab. A list of star maps for the Northern Hemisphere will come up. Under this list are three dots that when clicked will bring up a screen for maps of either the Equatorial Zone or the Southern Hemisphere.

Another resource for free star maps is the website www.skymaps.com.

[**Note:** Should these websites becomes unavailable, do a browser search for "free star maps" to find another resource.]

❷ On the Starmap site, the students will select the hemisphere for their location, the map version to be downloaded, and the time they are to view the stars (20 pm is 8:00 pm and 22 pm is 10:00 pm). Have them click the correct link to download the map. Have them print the map.

This site also has inexpensive apps for mobile devices: however, these apps locate the stars when the mobile device is pointed at them, so have the students first make observations by using the downloadable star maps. Other apps are also available and can be found with a browser search.

❸ Have the students make sure the map they've downloaded is for the correct hemisphere, then have them study the map. Read the comments on the left side of the map with your students. Note that many of the stars and constellations can be seen with the unaided eye, but some require a pair of binoculars or a telescope.

❹ On an evening that is clear of clouds, have your students go outside at the time recommended for the map they've downloaded. Have the students first spend some time just observing the stars.

If you live where it is difficult to see many stars because of artificial outdoor lighting, you may need to travel to a darker location.

❺ Have the students hold the map face down above their head so they can see it and the night sky. Then have them orient the map to the stars in the sky.

❻ Using the star map as a guide, have the students look for constellations as landmarks. Once they have found a few constellations, have them make their own star map by recording the constellations they see. On the map they create, have them note the magnitude of the stars, planets, or other objects and their location. Space is provided in the *Results* section for the map to be drawn.

Results

Space is provided for drawing a star map.

III. Conclusions

Have the students answer the questions and draw conclusions based on their observations.

IV. Why?

Read this section of the *Laboratory Notebook* with your students.
Discuss any questions that might come up.

V. Just For Fun

Have the students repeat the experiment in a month, making a second map. Than have them compare their two maps to determine whether the star locations have changed and to note any other observations they may have.

Experiment 20

Modeling Our Solar System

Materials Needed

- 8 objects of different sizes to represent the planets
- ruler (in centimeters)
- marking pen
- large flat surface for drawing—1 x 1 meter (3 x 3 feet), such as a large piece of cardboard or several sheets of construction paper
- large open space at least 3 meters (10 feet) square
- push pin
- piece of string one meter (3 feet) long
- additional objects of students' choice to represent asteroids, etc.

Objectives

In this experiment students will make a model of our solar system.

The objectives of this lesson are to have students:

- Observe the planetary orbits.
- Gain a basic understanding of distances between the planets.

Experiment

I. Think About It

Read this section of the *Laboratory Notebook* with your students.

Ask questions such as the following to guide open inquiry.

- *Why do you think the Sun and the planets around it are called a solar system?*
- *Do you think other planets could come to join our solar system? Why or why not?*
- *Do you think our solar system could join another solar system to make one big solar system? Why or why not?*
- *How accurate do you think planetary distance measurements are? Why?*
- *Do you think the Sun's gravitational field extends beyond our solar system? Why or why not?*
- *Why do you think the planets' orbits are almost circular?*
- *What do you think would happen if the planets' orbits were long ellipses? Why?*

II. Experiment 20: Modeling Our Solar System

Have the students read the entire experiment before writing an objective and a hypothesis.

Objective: Have the students think of an objective for this experiment (What will they be learning?).

Hypothesis: Have the students write a hypothesis. The hypothesis can restate the objective in a statement that can be proved or disproved by their experiment.

EXPERIMENT

❶ Help the students find 8 objects of about the right size to represent the different planets. They can use the illustration in the *Student Textbook* to see the relative sizes of the planets. Ask them how big the object representing Jupiter would need to be compared to the size of Mercury, etc.

❷-❸ Have the students use a marking pen to put a dot at the approximate center of the cardboard to represent the Sun. Then have them put a push pin securely in the spot they have marked and fasten one end of the piece of string to the push pin.

❹ Have them measure 10 cm from the push pin and put a mark there. Then they will take the marking pen, wrap the free end of the string around it so the marker tip is at the 10 cm mark when the string is tight, and draw a circle around the push pin at the 10 cm distance. This will represent Earth's orbital path.

❺ Have the students use the chart in the *Laboratory Notebook* to measure and draw the orbits for the first 5 planets (Mercury through Jupiter). They will be adding 4 orbits since they have already drawn Earth's.

❻ For the first 5 planets from the Sun, have the students place the objects they've chosen in the appropriate orbit for the planet being represented.

❼ For the last three orbits, have the students measure the distance of the orbit from the center and place the appropriate object at the orbital distance for the planet. Since these orbits are so far from the center, orbits don't need to be drawn for the outer three planets.

Results

Have the students analyze their model, comparing it to the illustration in the *Student Textbook*. What similarities and differences can they observe? Ask what other observations they can make about our solar system. Have them record their observations.

III. Conclusions

Have them draw conclusions based on the data they collected. How easy or difficult was it for them to create a model of the solar system? How did the different distances affect their model?

IV. Why?

Read this section of the *Laboratory Notebook* with your students.
Discuss any questions that might come up.

V. Just For Fun

Have the students add to their model by finding additional items to represent the objects in the Asteroid Belt, comets, spacecraft, etc. Have them note whether the objects are to scale.

Experiment 21

Discovering Life on Other Planets

Materials Needed

- imagination
- pencil
- colored pencils

Optional

- several sheets of blank paper

Objectives

In this experiment students will explore thought experiments by imagining what might be needed to travel to another solar system and look for life there.

The objectives of this lesson are to have students:

- Understand the value of thought experiments to scientific discovery.
- Think about what factors are required for life as we know it to exist in another solar system.

Experiment

I. Think About It

Read this section of the *Laboratory Notebook* with your students.

Ask questions such as the following to guide open inquiry.

- *Do you think a planet that supports life as we know it could be be either a Jupiter-like or an Earth-like planet? Why? If life were present on both types of planets, would it be the same type of life? Why?*
- *Do you think a planet that supports life would need to have liquid water? Why or why not?*
- *What do you think would be required for a planet to have liquid water?*
- *What distance from the parent star (sun) do you think a planet would need to be for it to support life as we know it?*
- *Do you think the chemistry of life on a planet outside our solar system could be different from Earth's carbon-based life? Do you think it could be silicon-based? Based on metals? Something else? Why or why not?*
- *If metal-based life is possible, could the exoplanet be closer or farther from the parent star than one that has carbon-based life? Why?*

II. Experiment 21: Discovering Life on Other Planets

Have the students read the entire experiment before beginning.

This is a different type of experiment for students to explore. The thought experiment is a very useful technique for creative inquiry.

Finding Life—A Thought Experiment

Encourage the students to use their imagination in thinking about what life on other planets and moons would require, how they would define life, and how they would go about looking for it. Have them think about what tools they might need to use for exploring other worlds.

IV. Why?

Read this section of the *Laboratory Notebook* with your students.
Discuss any questions that might come up.

V. Just For Fun

Have the students review their notes and make a diagram of the solar system they explored in their thought experiment. Have them indicate the different celestial bodies, make up names for them, and label them on their diagram.

Students will then continue their thought experiment by exploring ideas about what life might be like on an exoplanet. The questions that can be asked are endless and the objective is to have fun exploring the possibilities. By imagining what life could look like on an exoplanet, students will gain a deeper understanding of why life is possible on Earth.

Have the students write and/or draw as many details as possible about what life might be like on an exoplanet or its moon. Encourage them to use their imagination even if their ideas seem implausible.

Experiment 22

Working Together

Materials Needed

- computer with internet access
- materials as needed for project chosen by students
- blank paper or notebook

Objectives

In this experiment students will explore using the internet to participate in collaborative science experiments.

The objectives of this lesson are for students to:

- Experience how collaborative science works.
- Use collaborative science to gather and submit research data.

Experiment

I. Think About It

Read this section of the *Laboratory Notebook* with your students.

Ask questions such as the following to guide open inquiry.

- *Can you think of an experiment you have done that required collaborating with others? How did each person participate in the experiment?*
- *Do you think if you were doing a complicated experiment it might help to have collaborators? Why or why not?*
- *Do you think discoveries in science can be made more quickly if scientists collaborate on a project instead of working alone? Why or why not?*
- *Do you think a scientist might need to work alone on some experiments? Why or why not?*
- *Do you think you would prefer to do an experiment on your own or as a collaboration? Why? Do you think it might depend on the experiment? Why?*

II. Experiment 22: Working Together

Have the students read the entire experiment and choose a project before writing an objective and a hypothesis.

Objective: Have the students think of an objective for this experiment (What will they be learning?).

Hypothesis: Have the students think about what they might learn from doing science in a collaborative project. The hypothesis may depend on the project they select.

EXPERIMENT

1. Have your students go to the SciStarter website: http://scistarter.com/index.html

   Websites change from time to time. If SciStarter becomes unavailable, do a web search on *citizen science projects for kids* or a similar phrase to find another activity.

2. Students are asked to click on *Pick an Activity* which will list projects by location (e.g., on a walk, at home, in a car), or they can click on *Pick a Topic* which lists projects by the area of science being explored.

3. Have the students spend some time exploring the site and reviewing some of the experiments.

   Some recommended projects:

   - Measuring the Vitamin C in Food (Topic-->Food).
   - The Art of Cystallisation - a global experiment (Topic-->Chemistry).
   - Poo Power! Global Challenge (do a search on the website for Poo Power).
   - Drug discovery from your soil (Activity-->At Home-->).
   - Identify the Cloud (Activity-->On a hike-->).
   - A Project of the Day.

   Have them write down the names of experiments that look interesting to them and have them think about whether the experiment applies to your location and whether it is one they can do. Many of the experiments don't cost anything to perform, but some of them have associated costs for materials.

4. Help the students select a suitable project from their list. Students can do a search on the name they've written down to find the project again.

5. Have the students click on the *Get Started* button and follow the directions for signing up for the experiment and for downloading any instructional materials they will need. Help them gather any materials required for the experiment.

6. Have the students follow the instructions to perform the experiment, collect the data, and submit it to the experimental group.

Results

In the space provided have the students record the name of their project and the organization sponsoring it and describe what steps they followed, what data they collected, and how they think the data will be used. Have them record any other observations they can think of.

III. Conclusions

Have the students discuss what they learned about crowdsourced science by participating in this experiment.

IV. Why?

Read this section of the *Laboratory Notebook* with your students.

Discuss how using collaborative science could accelerate the discovery of new treatments for diseases and solutions for other science problems.

V. Just For Fun

Students are asked to make up their own collaborative experiment. Help them choose an experiment from the suggestions, or they can make up their own.

Have the students use separate sheets of paper or a notebook to write their own experiment based on the format of the other experiments in this book, including a title, an objective and hypothesis, materials list, list of participants, steps to be followed, results, and conclusions. Ask them to think about how they will make this a collaborative experiment. In what ways will others be involved? Who will be involved? Will each participant do the entire experiment or just certain parts? How will the data be handled?

When they have completed the experiment, in the *Conclusions* section have them record how well they think the experiment worked and how doing it as a collaboration was helpful and/ or not helpful. What did they learn? Did they learn more than they could have by doing it themselves?

Have the students fasten their completed experiment in their *Laboratory Notebook.*

More REAL SCIENCE-4-KIDS Books
by Rebecca W. Keller, PhD

Building Blocks Series yearlong study program — each Student Textbook has accompanying Laboratory Notebook, Teacher's Manual, Lesson Plan, Study Notebook, Quizzes, and Graphics Package

Exploring Science Book K (Activity Book)
Exploring Science Book 1
Exploring Science Book 2
Exploring Science Book 3
Exploring Science Book 4
Exploring Science Book 5
Exploring Science Book 6
Exploring Science Book 7
Exploring Science Book 8

Focus On Series unit study program — each title has a Student Textbook with accompanying Laboratory Notebook, Teacher's Manual, Lesson Plan, Study Notebook, Quizzes, and Graphics Package

Focus On Elementary Chemistry
Focus On Elementary Biology
Focus On Elementary Physics
Focus On Elementary Geology
Focus On Elementary Astronomy

Focus On Middle School Chemistry
Focus On Middle School Biology
Focus On Middle School Physics
Focus On Middle School Geology
Focus On Middle School Astronomy

Focus On High School Chemistry

Super Simple Science Experiments

21 Super Simple Chemistry Experiments
21 Super Simple Biology Experiments
21 Super Simple Physics Experiments
21 Super Simple Geology Experiments
21 Super Simple Astronomy Experiments
101 Super Simple Science Experiments

Note: A few titles may still be in production.

Gravitas Publications Inc.
www.gravitaspublications.com
www.realscience4kids.com